Praise for LAST OUT

LAST OUT is a raw emotional journey with no illusions. It is uncompromising, intense, and intimate, and for those open to see a different side of those who serve. Based upon true stories, LAST OUT more than suggests, there is never a "last one" out from the burden of war. The legacies continue through generations. LAST OUT aims and succeeds at showing that to those open to it.

David Segal, DC Metro Theater Arts

This is as good as theatre gets.

Dr. Anthony Chase, Theatre Critic

LAST OUT tells the real story of the American soldier... A more detailed and nuanced narrative about what it means to be a soldier. The highs and lows, the camaraderie and frustration.

Connecting Vets

Shines a light on a military story that Hollywood has left behind.

Salute Magazine

Without being politically divisive, the play shows the inner workings of our 18-year war in Iraq and Afghanistan from the standpoint of those who serve and are tasked with forming deep bonds with Afghan locals.

Noozhawk

This play is IT'S A WONDERFUL LIFE, with body armor.

Mike Rowe

LAST OUT is saving lives.

John Ondrasik, FIVE FOR FIGHTING

This play is as real as it gets.

Shawn Ryan

This play is a powerful tool for healing.

Tom Brokaw

LAST OUT

ELEGY OF A GREEN BERET

SCOTT MANN

IBIS BOOKS

LAST OUT: ELEGY OF A GREEN BERET

Published by Ibis Books, 2349 Hyde Park Street, Sarasota, FL, 34239.

Cover by Emma Brannon

First edition 2024

ISBN-13: 978-1-956672-25-1 (Paperback edition)

ISBN-13: 978-1-956672-26-8 (Ebook edition)

To all the Quiet Professionals and their Families...

This story is yours.

Foreword

As I write in my book, *Grateful American: A Journey from Self to Service*, I have had the profound honor to advocate around the world for the most amazing group of people I've had the pleasure to know: the men and women of the American Armed Forces and their families.

And as an actor and director, I've had the unique opportunity to express myself through theater. I've long held that theater and storytelling can provide a similar outlet for veterans and their families to heal while re-connecting to their civilian neighbors at a community level.

In 1984, a very special moment in my life came when I had the opportunity to direct *Tracers*, a play about the Vietnam War, as a major production at Steppenwolf Theater in Chicago. Conceived by John Difusco and written with the original cast of his fellow Vietnam veterans, *Tracers* was a transformative experience for these men as they brought their own stories to the stage, creating a play about war that helped to inform civilians and bring healing to them and the men and women who had served alongside them.

Hundreds, if not thousands of Vietnam Veterans saw our production during its run. Many of these veterans had felt shunned and neglected for years. At the conclusion of each performance, they would rush up onto the stage to hug and thank our cast. Their stories were finally being told. It was a powerful and emotional experience for me. *Tracers* burned into my soul the power of theater to heal the wounds of war through storytelling.

Little did I know, the groundbreaking production of *Tracers* planted another seed of healing and connection from a war that would come decades later.

Thirty-six years after *Tracers* was presented at Steppenwolf, retired Army Green Beret, Lieutenant Colonel Scott Mann was struggling, much like John Difusco, to find meaning in his post-war existence. As the war dragged on, he realized that most Americans were not connected to the long wars in Iraq and Afghanistan. He knew this disconnection between the protectors and protected did not bode well for our nation and the future of the next generation of warriors, including his oldest son who was on the path to join the Army.

Already wrestling with how he might tell the story from the stage, Scott came across the story of *Tracers* in my book *Grateful American*. Buoyed by the realization that combat veterans had indeed written and performed a story about the Vietnam War, he doubled down on his efforts to find a way to tell the story of his own generation's war; "a post 9-11 *Tracers*" in Scott's words.

The result was *Last Out*.

Scott and his team of veterans and military family members assembled a cast and production that toured 16 cities and brought the power of warrior storytelling to thousands. This story follows the journey of one Green Beret Sergeant, Danny Patton, and his family as they try to navigate the corrosive gears of a 20-year war. One of the things Scott has done with this play, beyond the combat realities of *Tracers*, is to toggle back and forth between the firebase where Danny operates, and his living room where his family tries to keep things moving forward.

Last Out has been performed for dozens of Gold Star Families, military families, veterans, and civilians. It was this blended audience that brought communities into the same room to talk openly about the hard subject of war, and to find a shared perspective while also re-distributing the emotional load from the shoulders of those few who were carrying it.

After the attacks of 9-11, I ramped up my support for our service members and their families. It became my life's mission to stand for them no matter what challenges they faced. As I visited our troops, toured with my band to lift them up, and lent my services to many military support organizations through speaking engagements and fund raisers, Scott and I barely missed each other on several occasions as he performed his play.

After the catastrophic withdrawal from Afghanistan, I felt the deep pain that so many Afghanistan veterans and military families were experiencing. It was shortly after this collapse that I had the opportunity to see *Last Out* for the first time. All the emotion from *Tracers*, and the pain I'd seen in so many Vietnam Veterans, came flooding back. Since the attacks of 9-11, I have personally been laser focused on supporting our military families, and it is the reason, in 2011, I created the Gary Sinise Foundation. Education of the wartime experience of our veterans is part of our mission, and Scott and I agreed that *Last Out* is the post 9-11 equivalent to *Tracers*. This continuation of warrior storytelling is a powerful way to educate our fellow citizens as to the cost of war, to bring healing to todays veteran, and to help prevent this generation from falling through the cracks as our Vietnam veterans had all those years ago.

Whether you are reading this script for entertainment or in preparation for a stage performance, what you hold in your hands are the stories of men and women who fought the longest war in American History. For twenty years, these warriors and their families endured the endless chaos of combat, deployment after deployment. It is a timeless story of the Forever War.

War stories heal the spirit and the mind. War stories validate. War stories inform. I saw it happen with *Tracers* years ago, and it is happening with *Last Out*. As the guns go silent from the longest war in U.S. history, the craft of warrior storytelling can bring a shared perspective to help us learn the lessons from the cauldron of combat. In our fast-paced society, where less than 1% of our population lives this back-and-forth wartime purgatory of being caught between ones living room and firebase, the disconnect between the 1% and the rest of us is clear. This play helps us to bridge that disconnect, as it lets us see into the lives of the men and women who fight our wars. *Last Out* is raw and it's real.

I am deeply honored to be on this journey of healing and connection with Scott... and with you.

You've heard the stories of the first in. This... is *Last Out*.

Gary Sinise

Preface

Stories—even the painful ones—heal our hearts, our brains, our communities, and our nation.

Last Out is one of those stories—a story of love, honor, and letting go.

Last Out was pulled from the experiences of the longest war in American history, a war mostly fought in the shadows. This play was first presented with a small cast of combat veterans and military-family members playing multiple roles. An elegy is a lament for the dead, therefore all scenes take place somewhere in the afterlife, spanning different times from Danny's youth and military career from May 1989 until his death in 2015.

Last Out is not an autobiography. But it is based on true stories of the men and women I served with, many of whom are no longer with us. This is a universal story to honor those who served in combat and the families we left behind.

Every story has a hero, and this one is no different. Our hero in *Last Out* is Lynn Patton, Danny's wife, the rock of the family, the military spouse. Lynn is inspired by Monty Mann, my wife, who has been on this mission with me for over 24 years. Our military spouses are a national treasure and it's time their story is told.

Danny's best friend in the play is Major Kenny Suggins. The inspiration for this character came from my "Ranger Buddy" Cliff Patterson, who was killed in the Pentagon on 9-11. A picture of us at my Granny's house on Christmas 1992 sat on the Wall of Honor during each performance of our first tour.

Military life is hard on a family, especially the kids. Wondering if mom and dad are okay and if they are coming home is part of the everyday reality for our military kids, just like it was for my three sons. It's bittersweet when they come to you quietly and say "I'm joining up," as my oldest son Cody did.

Green Berets are masters at building relationships with indigenous people. Many people don't realize just how closely we work with local tribes and clans around the world. Another hero in this story is Malik Jan, an Afghan tribal elder. This character is based on several real-world elders who mentored me through the years. So much wisdom and courage are contained within these quiet leaders.

Deployment *hellos* make you feel as if you're walking on air and the *goodbyes* make you feel like you have a hockey stick stuck in your throat. At some points in war, you lose track of which side of the ocean you're on and it all becomes a blur.

Throughout history, societies have used storytelling to help warriors integrate back into society, and *Last Out* follows in that tradition. The mission of this production is to validate the journey of our military veterans and their families while building genuine and well-informed understanding in the communities where they live.

Kit up, we roll out in five. It's time for *Last Out*.

Scott Mann

Tampa, Florida

PRODUCTION HISTORY

Last Out: Elegy of a Green Beret received its world premiere from The Heroes Journey (Monty Mann, Producer) in Tampa, FL, on November 10, 2018. The production then embarked on a sixteen-stop tour, February 2019-January 2020. The production was directed by Ame Livingston; with scenic design by Mark Hartley, costume design by Scott Mann, lighting design by William Glenn, and audio/sound design by Mark Prator, "Big Bob" Ballas, Wes Dearth, and Tony Leps (Red Front Door Studios). The stage manager was Bob Pritchard and the Assistant Stage Manager was Kelley Astyk. Filming was done by Kevin Lang. The Tour Manager was Kari Ellis, with logistics by Chris & Alliy Vetzel and Jimmy Klingensmith. The Production Manager was William Glenn, and Jamie Dunn handled Media. The cast was as follows:

DANNY PATTON: Scott Mann

OPERATOR 599, LYNN PATTON, INTELLIGENCE OFFICER ROBERTS, PENTAGON FIRE FIGHTER: Ame Livingston

OPERATOR 837, RICKY ADCOCK, JEREMY SMITH, SFAS INSTRUCTOR, SPECIAL FORCES MEDICAL INSTRUCTOR, ROBIN SAGE INSTRUCTOR, PENTAGON ARMY OFFICER, FREDDY, CAIDEN PATTON: Bryan Bachman

KENNY SUGGINS, MALIK JAN: Lenny Bruce

***

Last Out: Elegy of a Green Beret premiered as a film on May 31, 2021, on Amazon Prime, Hulu, Apple TV, and Google TV. Filming was done by Kevin Lang. The cast was as follows:

DANNY PATTON: Scott Mann

OPERATOR 599, LYNN PATTON, INTELLIGENCE OFFICER ROBERTS, PENTAGON FIRE FIGHTER: Ame Livingston

OPERATOR 837, RICKY ADCOCK, JEREMY SMITH, SFAS INSTRUCTOR, SPECIAL FORCES MEDICAL INSTRUCTOR, ROBIN SAGE INSTRUCTOR, PENTAGON ARMY OFFICER, FREDDY, CAIDEN PATTON: Bryan Bachman

KENNY SUGGINS, MALIK JAN: Lenny Bruce

***

Last Out: Elegy of a Green Beret received its second production from The Heroes Journey (Monty Mann, Producer) in Tampa, FL, on January 26th, 2023, with support from the Gary Sinise Foundation. The production then embarked on a nine-stop tour, January 2023-October 2023. The production was directed by Karl

Bury; with scenic design by Mark Hartley, costume design by Scott Mann, lighting design by Paul Hudson, and audio/sound design by Mark Prator, Ozzy Giron, Collin Hardesty, and Wes Dearth. The stage manager was Rebecca McBee and Katrina West. The Tour Manager was Megan Nemanic, and Assistant Tour Manager was Emma Brannon. The cast was as follows:

DANNY PATTON: Scott Mann

KENNY SUGGINS, MALIK JAN: Lenny Bruce

OPERATOR 599, LYNN PATTON, INTELLIGENCE OFFICER ROBERTS: Heather Corrigan

OPERATOR 837, JEREMY SMITH, SFAS INSTRUCTOR, GRADUATION PRESIDING OFFICER: Bryan Bachman and Tony Nevada

OPERATOR 595, STEVIE, RICKY ADCOCK, ROBIN SAGE INSTRUCTOR, FUNERAL PRESIDING OFFICER: Chris Vetzel

OPERATOR 726, FREDDY, CAIDEN PATTON: Cooper Mann

Fahim Fazli appeared as "Fahim" in the final tour stop. Gracyn Mix and Joshua Berduo were the understudies.

Characters (in order of appearance)

DANNY PATTON: male, 40s, Green Beret Senior Sergeant (Team Sergeant).

STEVIE: male, Danny's medic.

KENNY SUGGINS: male, 30s-40s, Lead Operator (see special note on Operators), Danny's original Detachment Commander, and one of his dearest friends.

OPERATORS 599, 837, 595, and 726: Valhalla Operators capable of shifting into different personalities and gifting recently departed warriors with powerful memories of their life. See special note on Operators.

SFAS INSTRUCTOR: Special Forces Assessment and Selection.

SPECIAL FORCES MEDICAL INSTRUCTOR

ROBIN SAGE INSTRUCTOR

GRADUATION PRESIDING OFFICER

LYNN PATTON: female, 40s, Danny's wife and best friend.

RICKY ADCOCK: male, Danny's Senior Engineer Sergeant and oldest friend on his Special Forces team.

PENTAGON ARMY OFFICER

FIREFIGHTER 1

FIREFIGHTER 2

ROBERTS: female, no-nonsense Military Intelligence Army Officer who provides targeting information to Danny and his team.

JEREMY SMITH: male, Special Forces Officer (Captain and then Lieutenant Colonel). Has a really big chip on his shoulder.

MALIK JAN: Afghan Pashtun Tribal Elder from Kara Khel Village.

FREDDY: male, Team Member.

CAIDEN PATTON: male, 20s, Danny's son.

FUNERAL PRESIDING OFFICER

ROLL CALL PRESIDING OFFICER

CAPTAIN ROBINSON'S VOICE

STAFF SERGEANT JOHNSON'S VOICE

SERGEANT FIRST CLASS CERVANTES'S VOICE

VOICE FROM OFFSTAGE

CASTING BREAKDOWN

ACTOR 1: male, plays DANNY PATTON

ACTOR 2: male, plays KENNY SUGGINS

ACTOR 3: female, plays OPERATOR 599, LYNN PATTON, ROBERTS, SPECIAL FORCES MEDICAL INSTRUCTOR, CAPTAIN ROBINSON'S VOICE

ACTOR 4: male, plays OPERATOR 837, JEREMY SMITH, SFAS INSTRUCTOR, GRADUATION PRESIDING OFFICER, PENTAGON ARMY OFFICER, STAFF SERGEANT JOHNSON'S VOICE, VOICE FROM OFFSTAGE

ACTOR 5: male, plays OPERATOR 595, STEVIE, RICKY ADCOCK, ROBIN SAGE INSTRUCTOR, FIREFIGHTER 1, MALIK JAN, FUNERAL PRESIDING OFFICER, SERGEANT FIRST CLASS CERVANTES'S VOICE

ACTOR 6: male, plays OPERATOR 726, CAIDEN PATTON, FREDDY, FIREFIGHTER 2, ROLL CALL PRESIDING OFFICER

CASTING NOTES

If at all possible, MALIK JAN should be played by an actor of appropriate ethnic background (Pashtun). If this casting proves impossible, then an actor who is of South or Central Asian descent, and who is comfortable portraying someone native to Afghanistan, should be used in the role.

Excepting the note above, all actors can be of any ethnicity/race. However, DANNY and KENNY should be of different ethnicities.

SPECIAL NOTE ON OPERATORS

There should be an outer ring struggle between the Operators as Danny struggles with his demons in the inner arena, and this struggle motors their persistent interruptions into Danny's world.

The Operators are from different wars in different times (see following notes) and therefore have different points of view and attitudes.

The Operators are badass, the elite of the elite. They're accustomed to operating alone but can operate in teams when necessary. However, they suffer no fools and don't necessarily take orders well when things start to go wrong.

Operators have a clear rank structure:

A: Probational. This is the entry-level for an Operator. Conditional engagement opportunities for warrior recovery. It's a kind of internship. A period to determine whether or not the individual is in possession of the necessary components to become an Operator. It is considered a lowly place/status and one you don't want to stay in for very long.

B: Journeyman. This is an Operator-in-training. They can engage in higher-level missions, but only with Operator oversight.

C: Operator. Past warriors certified to conduct solo engagements for recovery of warriors. Typically operate alone, but can also operate in team configurations under a Lead Operator.

D: Lead Operator. Extremely experienced Operators who are certified to lead recovery missions. They are usually brought in for the hardest cases of recovery like Danny Patton.

OPERATOR CHARACTER DESCRIPTIONS

KENNY SUGGINS

•Lead Operator.

•Former military occupation: US Army Special Forces officer.

•Killed in Action (KIA): 11 Sep 2001/Pentagon Washington, D. C./AQ attack on America.

•The quintessential Green Beret Officer.

•Good at building rapport and working with others.

•Extremely experienced in complex operations.

•Hundreds of recoveries under his belt.

•Patient. Kind.

•Doesn't tolerate anyone questioning his authority.

•Deeply connected to Danny and Lynn Patton.

OPERATOR 599

•Operator.

•Former military occupation: US Navy swift boat crew member

•KIA: 14 Jan 1968, Mekong Delta, Vietnam. (Firefight)

•Eager to become a Lead Operator.

•Has a very high mission recovery rate.

•Is by the book in her operational methods.

•Being groomed for Lead Operator status.

•Believes this mission, if successful, will lead to immediate promotion to Lead Operator.

OPERATOR 837

•Operator.

•Former military occupation: Army Air Corp bomber pilot (Brian).

•KIA: 12 Jan 1943, Dresden, Germany.

•Most experienced member of the team.

•Level-headed.

•Has seen just about everything in the warrior recovery arena.

•Driven to recover warriors as quickly as possible.

•Almost sucked into hell on several occasions and avoids this like the plague.

•Tolerates no reckless behavior from leaders or subordinates in this regard.

OPERATOR 595

•Journeyman.

•Former military occupation: US Army Infantryman.

•KIA: 16 Nov 2006, Ramadi, Iraq. Firefight

•Second most junior member on the team.

•High strung.

•A bit of a dickhead to subordinates.

•Very mission focused and competent, but abrasive.

•Impatient and dismissive with juniors.

•Believes juniors below him should be seen, not heard.

OPERATOR 726

•Probational Operator in Training.

•Former military occupation: Truck Driver.

•KIA: 23 Feb 2015 (Truck convoy IED, Kandahar, Afghanistan).

•Junior member on this mission.

•Thoughtful, persistent.

•Bumbling and clumsy at times and very nervous when around senior leaders.

•Desperate to prove himself to his operator teammates.

•Trying hard to get beyond his probation status on this mission, but the harder he tries, the more people don't seem to listen to him.

•Willing to jump into any scenario and show his worth as a potential Operator.

VALHALLA OPERATOR STANDING RULES OF ENGAGEMENT

1. Your mission is to recover warrior souls from the battlefield who are stuck, and escort them to Valhalla. At all costs.

2. Do not engage warriors whom you know. Recovery is rough business and friendships can compromise the hard tasks that must be done. Pass responsibility to another Operator.

3. You must recover lost souls immediately upon the stopping of their heart.

4. Do not allow warriors to "hang on" to residue from their life. This will prevent their Ascension.

5. Recovery must occur before hell launches its *Banyans* (evil-doers and mercenaries) and the warrior is condemned to frigid isolation.

6. Under no circumstances are you allowed to reveal inner workings of Valhalla or hell until the warrior "lets go."

7. Operators will report in and take all orders from Valhalla central (Odin).

8. If the warrior refuses to "let go" before hell's alarm for *Banyan* release goes off, Operators must abort and return to Valhalla immediately.

9. Operators must follow the Valhalla protocol when conducting warrior recovery:

•Protocol Level Alpha *(low threat).* Engage warrior directly and inform them you are there to escort them to Valhalla. You may expose them to non-combat memories.

•Protocol Level Bravo *(medium threat).* This indicates a growing level of warrior resistance to let go. Expose warrior to memories of combat and home life. Increase transition time between memories and deny long processing opportunities for individual memories. Keep the target disoriented.

•Protocol Level Charlie (*extremely high threat*). Threat of losing the warrior to hell's Banyans and eternal isolation is imminent. Employ all available memories, even the most painful ones, in order to shake the warrior loose and facilitate "letting go" and their Ascension to Valhalla. Be prepared to abort the warrior recovery when it becomes imminent for hell intervention. Failure to abort could result in Operators being dragged into hell with the warrior in question.

•*Note: Operator must request Valhalla authorization to move up protocol levels.*

10. Your mission is recovery. Be relentless. Do not let up. Constantly scan your target for opportunities and exploit memories by assuming identities of individuals with the most emotional impact on the target.

SETTING AND PRODUCTION NOTES

SETTING

Time

The entire action of the play happens in the time it takes Danny to bleed out on the battlefield. Scenes jump through memories and highlights of Danny's life and military career, ranging from May 1989 up through the *now* of the play: August 2015.

Place

This play takes place in Danny's mind. The space between life and death. Then, as the moments of Danny's life manifest, we jump through various locations, including Danny's home, his firebase in Afghanistan, villages, bars, battlefields, offices, Hell, and others.

PRODUCTION NOTES

The audience's imagination will do the majority of the work if you simply play the emotional truth. The notes below describe a highly effective approach, proven by the premiere production and first tour. Trust theatricality over literal realism.

Costumes

Danny should be the only character in actual fatigues. His kit should be as realistic as possible. All others can be in military cargo pants and black t-shirts. As Operators transform, they can simply add a piece or two to differentiate characters.

Scenic

An empty space, a handful of actor blocks, and the Wall of Honor.

The acting blocks are moved and rearranged to create various locations and furniture. All black works, but a gray/black/dark green camo paint treatment has also proven effective. Sandbags have added a nice touch.

The Wall of Honor is a multi-shelf system that includes, among other things, a folded American Flag, a large metal shard, a piece of flat granite, a picture of Kenny and Danny as Green Berets, memory bracelets, and other totems from Danny's life as a soldier. Feel free to include items and pictures from your cast and community. The original tour kept adding totems to the Wall of Honor as they manifested. Make it personal.

Four remotely operated red siren TIC (troops in contact) lights, one downstage left and the other downstage right.

For Danny's rifle, use a wooden cut-out, not a replica.

Start the play with a funeral memorial stand with desert boots downstage center.

Scenes should flow continuously into each other. No blackouts or rests. The momentum should slingshot and whipcrack. Scene and costume changes can and should be done in full view of the audience.

Music / SFX

The original SFX and music (including the various radio squelches, songs, and news reports) are available as part of performance licensing, or you may design your own soundscape. You are also welcome to do live sound, including having Operators read aloud the various off-stage voices and news reports, if that better fits your sound capabilities and artistic design.

Mental Health INtervention

In the course of the play, Lynn encourages Danny to attend meetings with The Heroes Journey, which offers healing through storytelling and narrative. You are encouraged to sub in the name of a mental health service or intervention group local to your production, and to include their contact info and other resources in your playbill and/or curtain speeches.

Language

Lastly, soldiers can be salty dogs. But their consistent use of obscenities and curses doesn't mean they aren't decent people at heart. It's their actions, not their f-bombs, that define them. However, if you feel your audience would absorb this story more readily if there weren't quite so much swearing, you have permission to clean up their language.

LAST OUT

ACT ONE

Scene 1

In darkness, we hear the sounds of combat. The TIC light activates and spills red light across the stage. The sound climaxes with a huge explosion.

Lights up. DANNY is dragged out by ACTORS 5 and 6. DANNY wears fatigues and body armor vest, and even mortally wounded he clings to his weapon. STEVIE does a combat assessment and tries to stop DANNY's bleeding.

Lights shift. DANNY rolls away and rises. Stevie continues—in mime—to desperately stop the bleeding as though DANNY were still on the ground before him. An erratic heartbeat pulses loudly as the sounds of combat continue.

DANNY sees his teammates in shock, not showing the typical discipline he expects of them in a firefight. He is amped up and pissed.

DANNY: What the hell are you guys doing?! Contact 3 o'clock, Captain. I need you to calm down, son. Get on the radio. We have an Eagle Down. Stevie, who are you working on, man? Stevie? Who are you...?

(DANNY looks over STEVIE'S shoulder and sees himself bleeding out. He recoils backward.)

DANNY: That's me?? Wait... that's me!! Atta boy, Stevie. Stop that bleeding. Just like I taught you. Don't give up. Don't let me die, man. It's not time yet. It's not time yet. It's not time yet!

(Everyone scatters as the heartbeat accelerates and gets louder. STEVIE continues to work as DANNY takes in his surroundings.)

DANNY: A roadside bomb? How did I miss it? You've been down that road a million times. Fuckin' cherry mistake! *(he looks around for a target for his rage)* Said Wali, you sorry son of a bitch, you got me, didn't you! I would give anything to rip your throat out! C'mon, Stevie! C'mon, man!

(The heartbeat bends and becomes a radio squelch.)

ODIN *(SFX: garbled radio transmission)*: Sergeant Patton... Valhalla!

DANNY: What the hell is this?

Scene 2

STEVIE disappears as the world shifts fully to "in-between." MAJOR KENNY SUGGINS enters and stares at DANNY. He has a walkie-talkie and is in this "in-between" place with DANNY but unseen by him. DANNY continues frantically exploring this new terrifying purgatory.

ODIN (SFX): **Garbled static transmission**

KENNY: This is Major Suggins. Over.

ODIN (SFX): **Garbled static transmission**

KENNY: Yes, sir. He just arrived.

DANNY: Lynn? Caiden?!

ODIN (SFX): **Garbled static transmission**

KENNY: No, sir. He's in post-combat mode. Suggins out.

(KENNY addresses DANNY.)

KENNY: Hey!

(DANNY wheels and levels his weapon on KENNY.)

DANNY: Kenny!?

(DANNY and KENNY embrace as only two lost brothers can.)

DANNY: Major Kenny Suggins!?

KENNY: Master Sergeant Danny Patton. They actually promoted you!?

DANNY: I saw the flames. You didn't make it out.

KENNY: That was a rough day.

DANNY: You think? Where am I?

KENNY: You're in between, man.

DANNY: In between?

KENNY: Yeah. I'm here to escort you.

DANNY: Escort me where?

KENNY: Valhalla.

DANNY: Valhalla?

KENNY: Valhalla!

DANNY: That shit ain't real, man.

KENNY: Yeah, it is.

DANNY: "Until Valhalla." You know what that is? That's some bullshit slogan for heaven to ease our pain when our friends die and we have to keep on living. That's what that is.

KENNY: No, it's real, bro. Don't I look real? It's up there in the clouds.

DANNY: Come on, man.

KENNY: No, look. Relax your eyes.

(DANNY reacts as he sees something beautiful and peaceful for a fleeting moment.)

KENNY: The Great Hall... more steak than you can eat! There's this Big shade tree of Glasir... cool, quiet stream underneath it. *(pause)* Nobody's shooting. *(They exchange a glance.)* You can stretch out on that soft green grass, and just rest. Come with me, Danny.

DANNY: Rest, huh? *(lost in the idea)* I don't know what that feels like anymore...

KENNY: You've earned it. Come with me.

DANNY: *(shaking off the fantasy)* It's not time for Valhalla.

KENNY: It's over. *(He points at the memorial stand with boots.)* Lay your weapon down.

DANNY: Fuck you. You want my weapon while Said Wali is still out there hunting my team?

KENNY: It's not your job anymore.

DANNY: What about my family? They are not ready for this. Where's Lynn? Caiden?!

(DANNY searches frantically for his family, for anyone he knows. KENNY moves a few steps away from DANNY and transmits. The erratic heartbeat pulses faintly, ominously.)

ODIN (SFX): **Garbled static transmission**

KENNY: No, sir. He's not ready. He's holding on tighter than a slip knot.

ODIN (SFX): **Garbled static transmission**

KENNY: He is not hopeless! I can reach him!

ODIN (SFX): **Garbled static transmission**

KENNY: Yes, sir. I understand his heart is giving out, but I need more time. Permission to start the Valhalla Protocol ASAP. I'll need you to send your best operators down here.

ODIN (SFX): **Garbled static transmission**

KENNY: Thank you, sir. Suggins out.

(OPERATORS 599, 837, and 595 walk on stage in military-style cargo pants and black t-shirts.)

OPERATOR 599: Major Suggins, I'm Valhalla Operator 599 and these are Operators 837 and 595. Where do you need us?

KENNY: Appreciate you coming down. Right there is your target.

OPERATOR 837: Holy shit.

KENNY: Yeah, he's stuck like hell.

OPERATOR 595: I haven't seen someone stuck this badly in a long time.

KENNY: Yeah, we're initiating Valhalla Protocol ALPHA.

OPERATOR 595: Roger that, sir.

KENNY: We have to become all of the people from Danny's life who shaped him and made his heart pump the most blood. We have to find out the reason why he's still holding on. We have to challenge him, we have to love him, we may even have to hurt him. But we have to force him to let go.

(OPERATOR 726 stumbles in, out of breath, with a big, excited grin.)

OPERATOR 726: Hey guys.

OPERATOR 599: You're late, 726.

OPERATOR 726: Yeah. Yeah, sorry.

KENNY: I didn't know we were getting four Operators.

OPERATOR 726: I'm probationary.

OPERATOR 595: Probationary? Why the hell do we have a probey on this mission?

OPERATOR 599: HQ directed it.

KENNY: OK, fine. 595, he's your responsibility.

OPERATOR 595: Fuck!

KENNY: Kid, this mission is touchy. Keep quiet and do what you are told.

OPERATOR 726: Yes sir.

OPERATOR 837: Why not just take him straight to combat and be done with it?

KENNY: Nah, his grip's too tight. Could be dangerous. I wanna try to shake him loose without taking him back to war.

OPERATOR 599: Do you traditionally go against standard retrieval protocols?

KENNY: For this one I am.

OPERATOR 837: You know the subject, sir?

KENNY: Master Sergeant Patton and I knew each other before I ascended—

OPERATOR 599: That is against protocol, sir.

OPERATOR 837: You are conflicted, sir. Perhaps another operator should take point on retrieval—

KENNY: Hey! That's the last time my command will be challenged. Do you understand?

(A beat. The OPERATORS glance at each other, then nod. The erratic heartbeat noticeably fades.)

OPERATOR 726: It's fading. We better get to it.

KENNY: Hit him hard upfront. *Pre*-combat. Got me?

ALL OPERATORS: Yes, sir.

KENNY: Let's get moving.

Scene 3

The heartbeat fades away and lights shift to highlight the Wall of Honor.

KENNY: Tell me about your wall.

DANNY: *(shaken from his trance; dryly)* I thought you left.

KENNY: I'm like a bad penny.

DANNY: That's my Wall of Honor! *(He examines it, lost in memories)* My son Caiden and I built it together after the war started.

KENNY: Wow. I'm glad you were a soldier and not a carpenter.

DANNY: Fuck you, Kenny. I wanted to help him understand why I had to be away so much. This wall held us together. At least, it was supposed to.

(DANNY pulls a metal shard off the wall.)

DANNY: My first kill in Afghanistan. I was brand new in-country. Kara Khel Village. Said Wali... that motherfucker... he was the target.

KENNY: Who is this Said Wali you keep talking about?

DANNY: You really don't know?

KENNY: No, man, I really don't.

DANNY: Biggest piece of shit on the planet. Taliban bomb maker.

KENNY: Bombmaker? That ain't exactly Osama bin Laden, bro.

DANNY: He is to me. We launched on him late one night after getting a tip from a local!

(DANNY slips into combat mode. Sudden light shift, sounds of infiltration. He is lost in a memory. KENNY attempts to pull him back.)

KENNY: Shit. Come back to me, Patton!

(DANNY moves in a crouch across the stage and engages a target close range.)

DANNY: Pop-Pop... Pop. Fuck you! *(crawls on his belly, sees a target and transmits)* This is ARANYA 08, enemy quick reaction force moving toward us from the north. Seven vehicles in the open. Request immediate close air support. Marking my location with Infra-Red strobe. You are cleared hot to drop. Over.

(An explosion. DANNY sees the carnage, leaps up, and lets out a war scream.)

DANNY:This is Aranya 08. Good kill! Out.

KENNY: *(dryly)* Congratulations.

DANNY: Thanks, man. It's not even a fair fight. That's fine with me... after what they did to us. That hunk of metal was all that was left of them. But, Wali wasn't in that convoy. He slipped away... like he always does. Fucking snake.

KENNY: Didn't take any Afghan commandos with you on the mission?

DANNY: No. So what?

KENNY: That's not the kind of Special Forces Mission I remember.

DANNY: Things change, man.

KENNY: Really? Did the color of your beret change?

DANNY: What the fuck is that supposed to mean?

KENNY: You know. Let me remind you where you come from, Green Beret.

SFX: squelch

Scene 4

Sounds of men running in formation, singing cadence, and clapping. KENNY moves DANNY toward a formation, and suddenly they are young candidates on the first day of Special Forces Selection.

SFAS INSTRUCTOR: Gentlemen. Welcome to Special Forces Assessment and Selection. For the next three weeks, you numbnuts are going to try out to become Green Berets. Out of six hundred candidates, only a handful of you will make the cut. This class represents an all-time low of the American gene pool. God help us.

(DANNY is at the position of Attention, smiling broadly. KENNY, also at attention, is next to him. Both men are sizing each other up. DANNY is arrogant as hell and pissing KENNY off.)

SFAS INSTRUCTOR: Roster #115. What the fuck are you smiling at?

DANNY: Nothing, Sergeant. I can't believe I'm finally here.

SFAS INSTRUCTOR: Me either, shit-for-brains. This life ain't for you. I can tell by looking at your scrawny ass.

DANNY: Oh, it's for me, Sergeant. I've wanted to do this since I was a kid.

SFAS INSTRUCTOR: Well shit. Isn't that sweet? Do you have any idea what you are about to get your dumb ass into? Special Forces operate hundreds of miles from nowhere, living in villages with no running water, speaking foreign languages surrounded by pissed-off tribes. Sound fun?

DANNY: I'll put up with anything if it lets me get after the enemy, Sergeant.

SFAS INSTRUCTOR: Fucking beautiful. *(Points to a stack of actor blocks where Operator 595 is sitting. Too late, 595 realizes he's been duped into playing the part of a jeep.)* You and your team are to push this jeep down this path until told to stop. Roger that?!

(DANNY and KENNY both swear under their breath, but a little too loudly.)

DANNY / KENNY: Fuuuck.

SFAS INSTRUCTOR: What's that, Rambo?

(DANNY jogs to the jeep.)

DANNY: Nothing, Sergeant. Wait, this jeep only has three wheels.

SFAS INSTRUCTOR: Oh, fuck! Oh, fuck! It only has three wheels? What'll we do? Look around. Adapt. Improvise. Do the best you can, Candidate. Roster #141!

KENNY: Yes, Sergeant!

SFAS INSTRUCTOR: Get in there and help "boy genius" figure this shit out.

(KENNY goes over. Both men glare at each other with contempt.)

KENNY: *(mocking)* I've wanted to do this since I was a little kid.

DANNY: Shut the fuck up and push.

(They start pushing. The jeep does not move.)

KENNY: Damn, Forrest Gump. Do you have any—I mean, is there even a little bit of ass in those baggy britches that can help push this jeep?

DANNY: *(stops pushing)* Did you call me Forrest Gump?

KENNY: *(also stops pushing)* Yeah. I did.

DANNY: Forrest Gump, huh? Tell you what, city boy. You quit running your mouth and we can finish this shit.

KENNY: Oh it's like that, huh?!

DANNY: It's just like that!

(DANNY and KENNY move face to face, about to fight.)

SFAS INSTRUCTOR: Hey! When you two jack-wagons are done slow dancing, somebody better push my fucking jeep or I'm putting you on the first truck home.

(DANNY and KENNY, mad mugging one another, resume pushing, slowly, as it weighs 3000 pounds.)

KENNY: Is that all you got? Why don't you put a little more of that scrawny, bumpkin ass into it?

DANNY: Why don't we build a sail and use your hot air to get this jeep across the finish line?

(They finish pushing. Both men are exhausted but exhilarated. DANNY extends his hand.)

DANNY: I'm Danny.

(KENNY shakes the offered hand.)

KENNY: I'm Kenny. Good to meet ya.

(Short radio squelch.)

KENNY: Well, Patton, selection is over. We made it. Next up: The Qualification Course. You go and become a lowly medic, while I bust my ass and become a Great Commander.

DANNY: Are you kidding me? I'm doing the hard part! You're going through the Officers' Course. All you have to do is learn to read a map and get a frontal lobotomy. I'll see ya when I see ya, brother.

(Man hug. KENNY walks off stage while DANNY sits.)

OPERATOR 726: Sir, I can become one of these instructors.

KENNY: 726, stay out of the way. Observe and learn.

OPERATOR 595: Hey shitbag, stay out of the way unless I tell you different. Get the fuck outta here.

OPERATOR 726: Yes sir.

(We hear goats braying throughout the next scene.)

SPECIAL FORCES MEDICAL INSTRUCTOR: Gentlemen, congratulations on making it through Selection. Welcome to the Special Forces Medical Course. The places your teams operate, there is no medical care. You dick-weeds have to do it all.

DANNY: Maybe I should have been an officer.

SPECIAL FORCES MEDICAL INSTRUCTOR: First intelligent thing I've heard you say yet, Patton. This shit is way too advanced for your simple ass. Special Forces Medics handle everything from battlefield tracheotomy, immunizing sheep, extracting an abscessed tooth on a village elder, or finding the right meds to treat your teammate's VD.

DANNY: *(raises hand)* Sergeant, can we get straight to the anti-VD part?

SPECIAL FORCES MEDICAL INSTRUCTOR: Oh, he's got jokes... We'll see how long you last in the Goat Lab, Patton.

DANNY: Goat Lab, Sergeant?

SPECIAL FORCES MEDICAL INSTRUCTOR: Live tissue training, fart-sack. To graduate as a Special Forces Medic, you must save the life of a goat that has been inflicted with an amputation, gunshot wound, severe burns, or other combat injury.

DANNY: Holy shit.

SPECIAL FORCES MEDICAL INSTRUCTOR: Sound cruel, cupcake? Maybe. But, the anatomy of a goat is remarkably similar to that of a human being. And when you're overcoming an obstructed airway on a four-month-old baby with third-degree burns or slapping a tourniquet on your buddy's leg as he begs to see his little girl again, you'll thank God for those goats. Their sacrifice makes us the most elite medics in the world. Now... if your goat dies—fuck you. You fail. Have a nice life. If your goat lives, congratu-fuckin-lations. You move on.

(Short radio squelch. KENNY enters.)

KENNY: Hola bitch hola!

DANNY: *(embraces KENNY)* Kenny! Good to see you, bro.

KENNY: I'm glad we're back together, brother. We're almost done.

DANNY: ...Yeah.

ROBIN SAGE INSTRUCTOR: Gentlemen, there are only 72 of you fuck-faces left. This is the Robin Sage Exercise: the final phase, and the largest guerilla warfare, role-player scenario in the world.

This is what makes Green Berets different. Tribal relationships. Subversion. Sabotage. Hand-to-hand combat. And that should keep you busy 'til lunchtime!

(ROBIN SAGE INSTRUCTOR exits. Once Danny sees the coast is clear he leans close to Kenny.)

DANNY: Hey man, I gotta talk to you. I am in trouble. Combat tactics and medical were no problem, but I've failed every event in this phase. They're getting ready to pull me out. I can feel it.

KENNY: That's because you're acting like Atilla the Hun. You're pissing off all the locals.

DANNY: Who?

KENNY: The role players, man.

DANNY: They are fucking morons. They don't know the first thing about combat ops.

KENNY: That's the whole point of this scenario, genius. You've got to build rapport with them.

DANNY: I'm not like you, Kenny. This rapport with the locals isn't my thing.

KENNY: Where's that coming from, man?

DANNY: My dad worked with tribes when he fought in Vietnam. Locals. Some of them turned on him in the middle of the night. Killed his best friend. He never got over it. So I'm about banging targets, not French kissing a bunch of wannabe guerillas.

KENNY: And how's that working out for you?

DANNY: *(sighs)* I hate it when you're right.

KENNY: Danny. Listen to me. Slow it down, okay? Do what I do. Connect with the locals. Teach *them* how to do the missions instead of trying to do it all yourself. It works, bro.

DANNY: All right. I'll do what I've got to do to get through this course, but I'm not ending up like my old man. You hear me?!

(SFX: radio squelch.)

GRADUATION PRESIDING OFFICER: Men of the 1st Special Forces Regiment. Congratulations. You made it. There are 1.4 million members of the U.S. Military, and only 6,500 Green Berets. Panamanian. Dominican. Black. White. Baptist. Muslim. Mystic. It doesn't matter. There is only one color on your team… Green!

(DANNY and KENNY fist bump without making eye contact.)

GRADUATION PRESIDING OFFICER: If there is even the slightest crack in your collective armor... one hairline fissure in your symbiotic, primary identity... Mission Fails. We all die. Look around you. This brotherhood starts here and lasts until your final resting place... in Valhalla. Gentlemen, you may don your Green Berets.

("Scotland the Brave" plays on the bagpipes. Operators applaud wildly. DANNY and KENNY pull their Green Berets from their cargo pockets, and put them on their heads. They stare into the horizon. Immovable. Long pause. SFX: short radio squelch.)

KENNY: Remember that moment, brother?

DANNY: Like it was yesterday. *(DANNY slowly removes and pockets his Beret.)* I loved every second.

KENNY: Love huh? Well hell, how about...?

(KENNY holds up a necklace. DANNY gasps and snatches it.)

DANNY: She wore this on the night we met.

Scene 5

Short radio squelch, then music plays. Use "Into the Atmosphere" by John Wesley, available as part of licensing.

DANNY: This song. This song played the first time we danced!!

KENNY: Not sure why she settled for your dumb ass, Patton.

DANNY: Me neither. I thought my team was all the family I'd ever need...

(LYNN enters and sits with a flip of her hair. Music continues to play.)

DANNY: ...until I met her. Hey Kenny, you know how in a fire-fight, everything is slow motion?

KENNY: Yeah.

DANNY: That's what it's like when you see your soulmate for the first time... hell, every time!

KENNY: Indeed. As I recall, she freight-trained your ass. We were in between missions in Fort Bragg, North Carolina, remember?

(DANNY follows KENNY into the club.)

DANNY: Uh-huh. You, Ricky, and me went out clubbing in downtown Fayetteville.

KENNY: Fayettenam!

(KENNY signals OPERATOR 595.)

OPERATOR 726: *(moving forward)* Can I—?

OPERATOR 595: *(stops 726)* Stay here.

(OPERATOR 595 moves into the scene as RICKY.)

RICKY: Yo! Fuckface!

(DANNY is thrilled to see RICKY alive.)

DANNY: Ricky!

RICKY: *(to Kenny)* Seriously? He brought us to the Palomino Club??

KENNY: Mr. Class, as always.

DANNY: Fuck you both very much.

LYNN: *(slightly buzzed from alcohol, addressing invisible suitors)* Oh, you're in the Army. Really? Wow, that's original.

DANNY: That accent, boys! Damn. I gotta look.

KENNY: Dang, man! She can't be from around here!

LYNN: No, I don't want to see your tattoo, GI Joe. *(Laughs out loud)*

RICKY: She's hot.

DANNY: Stay here, boys. I'm going in.

KENNY: He's a dead man.

(DANNY nervously moves toward LYNN, who has her back to him. He holds the necklace out.)

DANNY: Excuse me, Miss?

(LYNN and DANNY's eyes meet. Sparks fly. LYNN, trying to be coy, avoids DANNY's gaze. She smiles slightly to herself.)

DANNY: Is this yours?

LYNN: *(she reaches for it)* That's my St. Thomas—

(DANNY playfully pulls it away and finishes the line with her.)

LYNN / DANNY: —of the Cross.

DANNY: That's a good thing to wear during dark nights of the soul.

LYNN: That's my good luck charm.

DANNY: Well, then I should return it to its rightful owner. Miss?

LYNN: Lynn. Lynn Knauss.

DANNY: Well, Miss Lynn Knauss... may I?

LYNN: I guess.

(LYNN turns her back to DANNY, and he gently fastens the necklace around her neck.)

LYNN: And how can I repay your kindness, Mr...?

DANNY: Patton. Danny Patton.

(He smoothly spins her around and they start dancing.)

LYNN: Oh! Well, it's nice to meet you, Danny Patton. You sure don't waste any time, do you?

DANNY: Not when I see something that I need.

LYNN: *(rolls her eyes)* Something you need, huh?

(LYNN pulls away. RICKY comes to DANNY'S rescue, taps DANNY on the shoulder.)

RICKY: Hey Danny, introduce us to this Angel... Or you know what? Let me cut on in here. Hey girl, you like dancing? Check this.

(Ricky performs a ridiculous dance solo.)

DANNY: Don't ever do that again. I can't unsee that!

RICKY: Oh! I see what's going on here. No. NO! You don't call. You don't look at me the same.

DANNY: Oh, God.

RICKY: And the only sex we have anymore is hallway sex.

LYNN: And what is hallway sex?

DANNY: Lynn, don't ask him...

RICKY: It's when we pass each other in the hallway and we just yell "Fuck you!" at each other.

LYNN: *(smiling, she's loving this)* OK.

DANNY: Okay. Okay! Fine! Lynn, meet the degenerates from my A-Team.

LYNN: The A-Team, huh? Do you know Mr. T?

RICKY: Yes, we do.

(DANNY and RICKY look playfully at KENNY.)

KENNY: "I pity the fool!" You dumb motherfuckers.

DANNY: This painfully shy creature is Ricky. He is our engineer sergeant. He can build everything from roads to wells to bridges.

RICKY: Aaaaaand blow them the fuck up.

LYNN: Charming.

DANNY: And this handsome young devil is Captain Kenny Suggins. Our detachment commander.

KENNY: *(Kissing her hand. Whispers loudly.)* Run while you still can!

LYNN: Thanks for the advice. How long have you guys been together?

DANNY: Since the beginning. The three of us have been together since we started Special Forces.

KENNY: Yep. And now the Army, in all its infinite wisdom, put our dumb asses on the same team!

LYNN: Boy, the government didn't think that one through, did they? It's nice to meet you guys. How many more on your team?

DANNY: Twelve, total. These two are the only ones sober enough to come out tonight.

LYNN: Thank God for that. Well, since there aren't any Navy SEALS around to protect me tonight, I guess I will have to trust my safety to the Green Berets.

RICKY: Wow! She's a smartass, Danny.

DANNY: Yeah. I'm starting to see that.

LYNN: Smart enough to notice you. Come on, Danny Patton. Let's get out of here.

(RICKY and KENNY hoot and holler as DANNY and LYNN leave.)

Scene 6

The team and bar disappear. A short radio squelch transforms the music into the sound of a river.

LYNN: What river is this?

DANNY: It's the Cape Fear.

(LYNN takes DANNY'S hand as they sit side by side and watch the water.)

LYNN: It's beautiful. I can't remember the last time I sat on the tailgate of a truck.

DANNY: Well, we can't all be this refined.

LYNN: Look at all the fireflies! I love them. Did you know that they keep the darkness away?

DANNY: You know what I love? Listening to that river at night. Moving slow, powerful. Like it's searching for something.

LYNN: Maybe the ocean?

DANNY: Yeah. Maybe.

LYNN: So, what are you searching for, Danny Patton?

DANNY: You mean besides you?

LYNN: Blah. Nice line, soldier boy.

DANNY: Come on girl! You can't toss me a softball like that and not expect me to swing for the fence.

LYNN: Seriously... Why the Green Berets?

DANNY: I didn't want to be invisible anymore.

LYNN: Invisible?

DANNY: Yep. I was this scrawny runt in a little one-horse North Carolina Town. No one knew I existed.

LYNN: Oh, come on. What about your Mom and Dad?

DANNY: My Dad died when I was 14. Mom never really saw me after that.

LYNN: Was your dad in the Army as well?

DANNY: Yeah. For a few years. He was a Green Beret, too. He fought in Vietnam.

LYNN: *(smiling)* My Dad did too. He was a Marine.

DANNY: Yeah?! Those guys are cool. *(pause)* He never talked about it much. And I was always asking! Anyway, after he died, my mom and me were cleaning the attic. I found his war chest stuffed away.

LYNN: Oh, that is cool.

DANNY: His old team photos. His awards... including his silver star, and a beat up 1965 National Geographic.

LYNN: Wow.

DANNY: On the cover: "Army Special Forces in Vietnam." This little team, on their own, on the other side of the world... which seemed like a really great place to be at that point in my life.

LYNN: Yeah?

DANNY: Yeah.

LYNN: And you always knew that was what you wanted to do?

DANNY: A lot of bad things happened after Dad died. To Mom and me. Joining up seemed like a way to set things right. Maybe to finish what he started.

LYNN: Well, I like a man who knows what he wants. *(She stands and stretches languidly.)* I think I'm gonna take a dip in that river. Care to join me?

(LYNN begins to exit, goddess-like, and DANNY, after a stunned beat, leaps to follow.)

Scene 7

Slam to DANNY and KENNY. LYNN is across the stage.

KENNY: And it moved at Mach 3 from that moment on, didn't it, Patton?

DANNY: *(pulling a ring out of his pocket)* Check it out, man!

KENNY: So you're going to ask her, right? You're gonna pull the trigger?

DANNY: *(incredibly nervous)* I don't know.

KENNY: Are you crazy? You made it through all that bullshit training and now... you're going to pull back? You're going to pull back from that woman... who actually puts up with your shit?

DANNY: Dude. Not helping.

KENNY: I'm serious, man. That girl's good for you. Besides, if your dumb country ass can find somebody, I know good stuff's coming my way! Go get her.

(DANNY eyes the memorial and takes off his kit. OPERATORS watch intently... is this it? Is he laying his weapon down? DANNY then goes to the Wall of Honor and grounds his weapon and gear temporarily at its base, gathers himself, walks across the stage to LYNN, and takes her by the hand.)

DANNY: Lynn, can I talk to you?

LYNN: Always.

(DANNY drops to one knee.)

LYNN: Oh shit!

DANNY: Okay, that's not what I rehearsed!

LYNN: Oh my god, shit, god!

DANNY: Will you hush? Listen to me. I love you. I mean I love you so much it's... it's hard to breathe. This is gonna be a crazy ride. I mean, this shit's gonna be hard, girl. You hear me?

LYNN: Now you hush. Just tell me... is it worth it?

DANNY: This is big stuff we're gonna do, baby. Yeah. It's worth it.

LYNN: Well... then I guess the answer is yes. Yes!!

(He leaps up to embrace her but she stops him.)

LYNN: Just promise me one thing. When you stop loving it, if that day ever comes... we get off the ride... together. Deal?

DANNY: Deal, baby.

LYNN: Well?? You got a ring?

DANNY: Huh? A ring? *(smiles, then puts ring on her finger.)* It's not much.

LYNN: It's perfect. Like you.

(They kiss. Radio squelch and light shift. We hear a peal of wedding bells. DANNY picks her up.)

DANNY: Come here, Mrs. Patton! That was one helluva wedding.

LYNN: Best day of my life.

DANNY: Followed by a big slice of humble pie. Smallest damn base house in the entire US Army.

LYNN: Yes, but it's our tiny little house.

DANNY: It's fourteen square feet!

LYNN: Do you really have to deploy tomorrow, Danny Boy?

DANNY: I do, baby.

LYNN: Well, did you know that it's bad luck to leave your new bride without knocking her up first!

DANNY: These superstitions are really starting to pay off.

LYNN: So, you had better get your ass into that bedroom and start some pre-deployment activities, Sergeant Patton.

(They laugh and kiss. A radio squelch and light shift. A baby cries gently. KENNY picks up a swaddled infant and hands it gently to DANNY.)

KENNY: And then... Caiden.

DANNY: My shadow...

LYNN: I believe the tough Green Beret has met his match.

DANNY: Look at him, Lynn. He looks like a little possum. Listen to him breathe. I could listen to that for the rest of my life. What if I screw this up?

LYNN: You won't. Look at how he grabs your finger. He's strong. Like you.

DANNY: But he's way smarter. He's going to be a doctor... and a basketball player... and an engineer.

LYNN: You got this all figured out, huh? We'll see about that.

(Radio squelch and light shift.)

KENNY: *(takes the baby from DANNY and walks off stage)* Mach 3...

(DANNY moves downstage and sits.)

LYNN: What's it like without Kenny on the team?

DANNY: Weird. Hard to believe he's a Major now. Even crazier that he works in D.C. I miss him.

LYNN: I know you do. Did Caiden tell you what he wants for his birthday?

DANNY: Only a million times! The little Green Beret uniform from the PX. Just like mine. He says he's finally five and old enough to wear it! How funny is that? Check this out.

(DANNY stands and gathers his kit and rifle.)

DANNY: I even made him his own wooden rifle. *(holds up his weapon)*

LYNN: That's sweet. But I told him we can't afford it this year, so we'll save up for his next birthday.

DANNY: What? Not my little Possum. He's getting that uniform, Lynn.

LYNN: Sweetie... I want him to have it too, but we can barely pay the electric bill this month. Where are we gonna find the money for that uniform?

DANNY: *(proudly hands over cash to LYNN)* Here.

LYNN: What's this?

DANNY: I hocked the microwave. *(a sheepish grin)* Guess I should have told you.

LYNN: *(shaking her head but smiling and kissing him)* I love you, Danny Boy.

Scene 8

Slam to DANNY and KENNY sitting together, facing in different directions.

DANNY: Thanks, brother. Thanks for showing me all of this.

KENNY: You're welcome, man. It's why I'm here.

(We hear the faint erratic heartbeat. DANNY notices it.)

DANNY: What is that?

KENNY: That's you.

(Stevie becomes visible, still giving CPR.)

DANNY: Stevie's not gonna save me, is he?

KENNY: No, man. He's doing his best. But no.

DANNY: Jesus.

KENNY: Yeah. And we don't want to be here when your heart stops beating.

DANNY: What do you mean?

KENNY: Hell. Face-to-face with your darkest fears, clinging to your regrets, alone and drowning, for eternity. *(rises)* But I wasn't gonna let that happen. You ready?

(KENNY extends his hand to DANNY.)

DANNY: I think so.

(DANNY takes KENNY'S hand and stands.)

KENNY: Good. Let's go. Time's short.

DANNY: *(standing to put on his kit, but then lingering)* Okay, it's just... something doesn't feel right. Something...

KENNY: Danny, let's go, man!

(TIC light starts to flash.)

DANNY: That light. That fucking light! That's a Troops in Contact light. Every time that thing flashes in a headquarters, blood gets spilled.

KENNY: Danny!

DANNY: A fight's coming, Kenny. A big fucking fight!

KENNY: Danny, come on... Let's go, man! Right now!

DANNY: Kenny, there's something I have to do. There's something...

(Explosion. Chaos. KENNY is thrown back from DANNY, who has conjured 9/11. Darkness, except for TIC light.)

Scene 9

OPERATOR 599 rushes in and helps KENNY to his feet. DANNY picks up a piece of granite and sits.

KENNY: I'm fine, dammit. Get in there!

(TIC Light off. OPERATOR 599 steps forward as LYNN.)

LYNN: There you are. Jesus. I was so worried about you, baby. Look at you. Your uniform's a mess. How long have you been sitting out in this driveway?

DANNY: I dunno. A while, I guess. On the train ride home from DC… hundreds of American flags zipping by on the front porches. Most of 'em brand new. It'll be yellow ribbons real soon.

LYNN: Where did you get that rock, baby?

DANNY: It's granite. From the Pentagon.

LYNN: Danny…

DANNY: After my records review, I stopped by to see Kenny.

(Radio squelch and light shift as DANNY flashes back to the Pentagon. OPERATORS move rapidly in the background. Fear and confusion everywhere. DANNY approaches a PENTAGON ARMY OFFICER.)

DANNY: Excuse me, sir. What's going on?

PENTAGON ARMY OFFICER: Two planes hit the World Trade Center thirty minutes ago.

DANNY: Jesus!

PENTAGON ARMY OFFICER: What do you need?

DANNY: I'm looking for the Army Budget Office. Major Suggins.

PENTAGON ARMY OFFICER: Kenny Suggins? What you'll do is go down this hall, turn left—

(Another explosion. Lights flicker. DANNY and the OFFICER are thrown to the ground. Pandemonium. DANNY'S ears are ringing and he's choking on smoke. The OFFICER moans.)

DANNY: You're gonna be okay, sir. You got a nasty knock on the head, but you'll be fine. Medics are on the way.

PENTAGON ARMY OFFICER: *(reaching up)* Don't leave me!

(DANNY surveys the carnage. Distant sirens. Two FIREFIGHTERS rush by.)

DANNY: Hey, you have a man down over there. What the fuck happened?

FIREFIGHTER 1: An airplane hit the South wall.

DANNY: *(looks that way, realizes, then starts to rush toward the smoke)* That's Kenny's office!

FIREFIGHTER 2: *(grabs Danny)* What are you doing?! You can't go in there!

DANNY: Kenny!

FIREFIGHTER 2: Listen to me. No one made it out from there. They're gone, okay? They're all gone... I'm sorry.

(Another explosion. DANNY fights through the firefighters and drops to his knees. The OPERATORS look at the carnage, aghast.)

KENNY: It's ok, brother. You did everything you could.

DANNY: Fuck that, man. *(he stands)* They're gonna pay for this.

KENNY: Keep your head, Danny. Gotta be smart. Keep the long view on this.

DANNY: Long view my ass.

KENNY: You can't stay in between like this, Danny.

DANNY: Valhalla can wait!

KENNY: No, it can't! And it won't!! You've got to let go. If you're still here when your heart gives out, you'll be condemned to a place of unparalleled suffering. A barren wasteland, where you will be faced with all the wrongs you've done. And... you can never look away. But worst of all, it's cold, man. Real fucking cold.

(The OPERATORS are startled at KENNY breaking protocol.)

DANNY: Bullshit. I'm gonna do things my way now. My way!

(DANNY starts gearing up. The OPERATORS are pissed. OPERATOR 837 speaks to OPERATOR 599.)

OPERATOR 837: Suggins is losing control. His relationship with the subject is compromising the mission.

OPERATOR 726: Can I suggest something?

OPERATOR 837: Not now.

OPERATOR 599: *(addressing KENNY)* That information on Hell is classified, Sir. And you know that.

KENNY: I know what I'm doing. I'm going to save him. Tear his guts out.

Scene 10

Lights shift. DANNY is lost in vengeance. A clock ticks, slow and maddening. DANNY moves to LYNN who is sitting on the couch. They sit. They sit. The clock ticks. They sit.

LYNN: This is driving me crazy. I feel like I'm waiting for an execution. *(pause)* It's different this time, isn't it?

DANNY: Yeah. This deployment is different. But, it's the right thing to do. It's what I have to do... for Kenny.

LYNN: I've been going over it in my head... how it'll go at the airfield. The signs, the ribbons, and how I'll put on that bullshit brave face, and say all the bullshit things like how "it'll be over before we know it."

DANNY: Yeah. I really suck at this part.

LYNN: Then, soon as you boys go wheels up, I'll be kicking myself for all the wasted time I spent looking at that damn clock instead of just being with you when I had the chance. So, c'mon Danny Boy. *(suddenly more OPERATOR than LYNN)* Be with me. Let it go.

(DANNY considers, and then deflects.)

DANNY: Caiden sure is quiet. That damn soldier uniform is all he wears. We gotta wash that thing at some point.

LYNN: Like Father like son. Hey. *(she points to his wrist)* Is that Caiden's silly band?

DANNY: Uh-huh.

LYNN: The baseball player silly band?

DANNY: Yep.

LYNN: That's his favorite. The only thing he wears more than that soldier uniform is that silly band. He never takes it off.

DANNY: I know.

LYNN: He gave it to you?

DANNY: At breakfast, while you were still sleeping. He's staring into his cereal, not talking. Before I can ask him "what's wrong" he slips it onto my wrist. Whispers that "it's magic."

LYNN: Magic?

DANNY: Yeah. He says it will keep me safe… and bring me back to him.

LYNN: What else did he say?

DANNY: He made me promise to always wear it when I'm over there. Can you imagine that? Trying to keep this on with all the dickhead officers over there??

LYNN: *(deadly serious)* He's right.

DANNY: What?

LYNN: Don't you take it off while you are over there, you hear me?

DANNY: Lynn... look, I get the superstitions, but—

LYNN: Promise me!

DANNY: Okay. I promise. *(clock chimes)* Now we gotta head to the airfield. It's time.

(They stand. We hear the sound of jet engines spinning up.)

DANNY: That's my ride.

LYNN: (bravely) There she is.

DANNY / LYNN: I'll see you when I see you, but if I don't, know that I love you.

(DANNY and LYNN embrace. We hear the erratic heartbeat. LYNN looks back at KENNY, frustrated.)

OPERATOR 599: Sir, we are going to lose him.

OPERATOR 837: We need to take him to war, Sir.

(KENNY struggles with the truth. He pulls out his walkie-talkie.)

KENNY: This is Suggins. Permission to escalate to Protocol Bravo.

ODIN (SFX): (*Offstage radio transmission*) #%^&*! Valhalla $%^&

KENNY: Yes, I KNOW he's bleeding out. I'm tracking his heartbeat, Sir. Suggins out. *(he pockets his walkie-talkie)* Operators. Take him to war!

Scene 11

Radio squelch takes out the heartbeat, and lights shift. We hear an Islamic call to prayer.

RICKY: This place is crazy man. Like *RoboCop* had a baby with *Last of the Mohicans.* I just saw a camel spider outside bigger than my damn head.

DANNY: I don't care if we're throwing spears, Ricky. I just want to start hunting.

RICKY: Well, settle in, brother. This is gonna be a long war. You see all the shit they're building on this base? Harley Davidson Shop? TGIFridays!? This ain't no short-term gig.

DANNY: Wait, Ricky Adcock bitching about Harley shops and too much fast food? This place *is* crazy!

RICKY: No, man. This is what the Soviets did. They built up the same way... and they got their asses kicked.

(Short radio squelch. Slam to RICKY and DANNY in an orientation/briefing.)

ROBERTS: Gentlemen, welcome to Kara Khel firebase. I'm Roberts, your targeting officer for this deployment. Your firebase is adjacent to the village of Kara Khel. A worn-down farming community of about 1200 people. Forty years of drought, hard times, and famine. The local Taliban run the place. I'm not sure who's more dangerous... the insurgents, or the Afghan Police and government officials who are supposed to be on our side.

RICKY: What's the difference? They all hate us.

DANNY: Fine with me. I'm not here to make friends with the Afghans. Forget winning hearts and minds, let's kick some ass!

ROBERTS: Don't get your hopes too high, Patton. They call Afghanistan the Graveyard of Empires for a reason. I've been working intel for a long time, and there is no enemy like this insurgent.

DANNY: Bullshit. These guys are amateurs.

ROBERTS: Think so? You're not going up against fair-weather al Qa'ida foreign fighters. These are Taliban, hard-core local insurgents. They've been ambushing outsiders like you from the same ridgelines for centuries. They hand down family ambush sites from generation to generation, like your family hands you down your grandfather's pocket watch.

RICKY: That's how I got my Packers season tickets, dude.

ROBERTS: Don't look for help from the village either. A lot of the locals here, like that elder Malik Jan, they fear the Taliban way more than they do us. The Taliban Commander, Said Wali... he's got serious game. Do not underestimate him.

DANNY: I ain't interested in old men and farmers. Whaddya know about Wali?

ROBERTS: He's a player. Senior Taliban Council Member. He helped provide security while bin Laden's cronies made their plans. Now he has his foot on every throat in Khara Khel Village.

DANNY: So we gotta take him out. He'll be our first "jackpot."

ROBERTS: Easy cowboy. Every mission launched against Wali by every team before you has ended up with two things: "Dry Hole," and at least one casket draped with the Stars and Stripes.

DANNY: Not us... we'll get his ass.

(ROBERTS exits. Short radio squelch. Slam to DANNY and RICKY breaking down their gear, eating, waiting, or some kind of deployment zone business.)

RICKY: You hear about our new Captain?

DANNY: Smith? Nothing good. Why in the hell are we getting a new detachment commander right after arriving in-country?

RICKY: Cuz his Daddy's a general. I don't trust him, man.

DANNY: Yeah.

RICKY: I heard he lost his brother in the towers and now he's obsessed with racking up a body count.

DANNY: As long as he's targeting Wali, I'm in. A lot of us lost people that day.

RICKY: It's not a fucking scorecard, man.

DANNY: It is now. For Kenny, remember?

(SFX: short radio squelch.)

Scene 12

OPERATOR 726 steps in eagerly and addresses OPERATOR 595.

OPERATOR 726: Sir, I have an idea. If we—

(OPERATOR 595 holds up his hand, cutting off 726.)

OPERATOR 595: We're supporting 837. No talking.

(Short radio squelch. Slam to SMITH addressing the team. RICKY and OPERATOR 726 join DANNY in the team area.)

SMITH: Listen up. My name is Captain Smith. I'm gonna make this short and sweet. For years, the Special Forces motto has been *De Oprresso Liber*, "Free the Oppressed." Fuck that. Our new motto is "Punish the Guilty."

DANNY: *(preps gear)* That's what I'm talking about! Scalps on the barn, baby!

SMITH: *(preps gear)* It starts with that fucker, Said Wali—and anyone who gets in the way of me fucking him up!

DANNY: Yes, sir!

(TIC light starts to spin.)

ROBERTS: We have intel on Wali. The early warning network is thick. Every bad guy in the village will know you're coming. If you want the slightest chance of surprise, you'll need to land on the X: right on top of the enemy compound.

(Rotors and a hard thud landing sound.

DANNY: Stack up!

DANNY, SMITH, RICKY, and OPERATOR 726 line up in a stack outside a door. Offstage Breach Explosion. The team flows through the door and locks down the room. Sounds and movement of assault.)

DANNY: Bushmaster 6, this is Aranya 08... Dry hole. I repeat, dry hole.

(TIC light off.)

SMITH: FUCK! I want JACKPOTS, Motherfuckers! You hear me? I want Said Wali.

(RICKY and OPERATOR 726 exit pissed off.)

DANNY: We'll get him, Sir.

(TIC light starts up again.)

ROBERTS: We just got another intel update that Said Wali is shaking down some locals over in Khara Khel Village!

SMITH: How solid is the intel this time, Roberts?

ROBERTS: Not great. Only one source... not the most reliable one, either.

DANNY: Sir, the rest of the team is out on patrol.

SMITH: So what? Patton, you and I are gonna check this out. Get me some air support lined up.

DANNY: Captain, why don't we call the rest of the team in? If Wali's there, this could turn into a big hit.

SMITH: No time for that. I can handle Wali myself. Kit up, and let's roll out.

(TIC light keeps spinning.)

DANNY: Yes, Sir.

(DANNY and SMITH don their gear, move around the stage, take a knee. SMITH looks out over the audience.)

DANNY: *(in a tactical whisper)* Sir, air support on station. B-52 Bomber. 20 bombs in the rack. They are requesting permission to drop.

SMITH: I'll take it from here.

DANNY: *(beat)* Yes, Sir.

(DANNY reluctantly hands over the handset.)

SMITH: I'm finally gonna get this fucker.

DANNY: Sir. We don't have a good enough lock on Wali's location to call in a pre-emptive airstrike.

SMITH: Sergeant Patton. Do your job and focus on pulling security. *(into headset)* Red Ryder, this is Aranya 06... verifying grid coordinates. Stand by for my command.

DANNY: Sir, let's get closer and flush him out. Or talk to some of the locals and see if they've seen him.

SMITH: Patton, shut your fucking mouth.

DANNY: Sir, listen to me. Listen to me! This is a MISTAKE! You can't drop a bomb using a shitty piece of intel.

SMITH: If I want your opinion, I'll give it to you, Patton. Question my command again, and I'll rip that Special Forces tab off your shoulder and launch your ass back to Bragg for court-martial. Now pull security. *(into headset)* Red Rider, Red Rider, this is Aranya 06, you are cleared hot to drop, over.

(SOUND of distant explosions. TIC light off. Time stops.)

DANNY: Twenty-five! This dumb fucker killed 25 women and children in that drop. Said Wali was never even there. Set us back a year in that province, turned the Afghan president against us, recruited who knows how many more Taliban.

KENNY: So what did you do?

DANNY: There was an investigation into the airstrike. I told the Battalion Commander my version of events, that I tried to call Captain Smith off. They threw him off the team, and he accused me of ratting him out.

(SMITH glares at DANNY and exits.)

KENNY: Fuck. Target on your back.

DANNY: Big time. But as pissed as I was at Smith, I still blame Wali. He stood guard over those bastards as they planned your death. Fucker haunts my dreams, Kenny. Every time I rotate home, I feel him getting stronger while I sit here watching "America's Got Talent." I miss the fight. I have to get back there.

KENNY: But you just deployed home from your first Afghan tour, man.

Scene 13

Short radio squelch. Slam to DANNY in the hangar waiting to be picked up. We hear Army Band music in the background.

LYNN: *(entering on the run)* There's my soldier boy.

DANNY: Come here, girl!

(They embrace.)

LYNN: Oh, I missed you.

DANNY: Let me look at you.

LYNN: Here I am. Wanna see more?

DANNY: Definitely!

LYNN: Come on. Get in the car, baby. Let's go home.

(LYNN and DANNY sit in the car. DANNY looks in the back seat.)

LYNN: Is he asleep?

DANNY: Yeah, he's sleeping.

LYNN: You okay, baby? You with me?

DANNY: Yeah. I'm good. It's just...

LYNN: Just what, baby?

DANNY: Caiden wouldn't even hug me at the airfield. Did you see the way he was hiding behind your leg?

LYNN: Well, you've been gone a whole year. He's only six-and-a-half. Give it some time.

DANNY: (muttering) Time.

LYNN: Look at all the yellow ribbons.

(DANNY is bothered and anxious as he gazes out the car window.)

LYNN: Come on. Where are you in that head of yours?

DANNY: Everything's so... so... *clean*. So goddamn bright and shiny. It's offensive!

LYNN: Oh.

DANNY: It's like... four days ago, I was in a firefight, and out of ammo. But tomorrow morning, I'll be in the kitchen, and out of coffee creamer. What do you do with that?

LYNN: Mmmmm.

(LYNN makes a turn.)

DANNY: Where are you going?

LYNN: I gotta stop by the market.

DANNY: This late? Why?

(She glances at him.)

LYNN: Coffee creamer.

(They laugh.)

DANNY: She makes my blood warm again, Kenny. But I've gotta get back before it's too late.

KENNY: It's already too late.

(SFX: short squelch. KENNY gives a hand gesture command to LYNN. Light shift and sound of bird chirping. LYNN moves to the window. DANNY walks up behind her and wraps his arms around her waist.)

DANNY: Morning, baby.

LYNN: Morning.

DANNY: *(hands LYNN her coffee)* Your red bird back again?

LYNN: Yep. He shows up at the window every morning, just for me. Every morning. It's like he knows I need that lucky sign to start my day.

DANNY: And what do you do if he doesn't show up? Back to bed?

LYNN: You know something, mister? We all find what we need to get through this war. I don't talk about the fact that your only good luck charm is a gun. You leave my red bird alone okay?

DANNY: Sorry baby.

LYNN: Umm-hmm. You and Caiden done putting stuff on your little wall?

DANNY: Excuse me?! That is the Wall of Honor!

LYNN: *(chuckling)* Oh, pardon me.

DANNY: He wanted to know the story of every single thing I brought back. Even the hunk of metal. It's hard to bring that stuff back sometimes

LYNN: So what did you say?

DANNY: I changed the subject to the silly band he gave me.

LYNN: Good call.

DANNY: I told him we'd put it in its own special spot on the Wall every time I come back from deployment... and I'd put it on whenever I have to go back over. Except...

LYNN: Except he had no idea you were going back over.

DANNY: Yeah. He asked why I had to go back.

LYNN: And?

DANNY: I told him I didn't get all the bad guys.

LYNN: What did he say to that?

DANNY: *(pause)* He said he was going to grow up quick so that he could go over there and help me fight them.

(LYNN tries to process this. She looks away and nods.)

DANNY: I know. I know. He should be building tree forts. Not planning for Afghanistan. I'm going to finish this.

LYNN: How?

DANNY: I need to get back to being the kind of Green Beret that Kenny was. He always knew how to work with the locals. There's this little village near our firebase called Kara Khel. This old man, Malik Jan, is the elder. Powerful dude. He can't stand us, but if we could get him alongside us... well, that's the only way to win this thing.

LYNN: Hmmm...

DANNY: Ricky's right. It's gonna take time. It'll mean more deployments. Lots more. It won't be easy.

LYNN: Yeah, cause this has been such a cakewalk so far.

(DANNY looks away hurt.)

LYNN: Hey, I know that you can do it, baby. There is no doubt in my mind. *(she kisses him; they stay forehead to forehead)* But I think it's my turn to hock the microwave.

DANNY: What? Why?

LYNN: With that many more deployments, I gotta invest in a better vibrator.

(They laugh and kiss. LYNN exits.)

KENNY: How many deployments, Danny?

DANNY: Can't even remember. But I did write Lynn 710 letters.

KENNY: One before each mission. Just in case.

DANNY: Yep. And I got to rip up every single one.

KENNY: Except number 710.

(DANNY startles.)

Scene 14

Short radio squelch. Slam to Afghanistan. Sounds of Islamic call to prayer.

RICKY: *Another* Afghanistan deployment, asshole?! Why in the name of God did they promote you to team sergeant?

DANNY: Here we go.

RICKY: Can't be talent, otherwise they would've promoted me.

DANNY: I couldn't agree more.

RICKY: I know! I bet it's because you're *(whispers)* half-Ghan now.

DANNY: Hilarious, Ricky.

RICKY: Why else would you keep bringing us poor souls back here, man? Honestly, I don't know why I stay with you.

DANNY: It's probably the way I sing you to sleep at night.

RICKY: Green Berets are fragile creatures, Danny. We need variety. Occasional refinement. My levels run deep, bro.

DANNY: Oh, your levels run deep alright.

RICKY: Hey! How about you get us one of those nice cushy embassy gigs next, huh?

DANNY: *(smiling and shaking his head)* How about the beatings will continue until morale improves?

RICKY: How many more years are we gonna keep coming back to this same shithole village? My breath smells like okra. My ass smells like goat. My farts have fucking sand in them!

DANNY: Dude, you gotta take a bath.

RICKY: *(thoughtfully)* There's showers in this village?

DANNY: Look, as soon as we get Malik Jan and the elders to stand up on their own we'll be able to smoke Wali's ass. Done.

RICKY: Done? Really?

DANNY: Well, actually more like rinse and repeat. We still have about two dozen villages left after this one. Then—we are done.

RICKY: Hey, what about Caiden? He's just about old enough. When are you gonna bring him in?

DANNY: *(menacing)* Don't you ever say that again. *(starts to walk away)*

RICKY: *(suddenly serious)* Hey, I wanted to talk to you about something... The boys have been grumbling since Jason and Axel were hit last week.

DANNY: Yeah, I know. Axel... That dog found thirteen improvised explosives in one week.

RICKY: That beautiful black lab saved us all from getting blown to hell, man. Any word on Jason?

DANNY: Yeah. He's at Fort Sam. Headquarters said they had to amputate the right leg.

RICKY: Fuck. That sucks, man.

(RICKY is in deep pain and guilt. DANNY clocks it. Tries to comfort his friend.)

DANNY: There was nothing you could have done.

RICKY: Yeah, but maybe...

DANNY: Maybe what?

RICKY: Well, maybe the people who planted that bomb are the same ones we're trying to protect, Danny.

DANNY: They are not all bad. We just need to help them stand up.

RICKY: And how many of us die in the process?

(TIC light on.)

ROBERTS: We have intel coming in that Said Wali is coming into town tonight... gonna crash at his brother's house.

DANNY: Thanks, Roberts. *(to Ricky)* Okay! I wanna hit that motherfucker up close and personal. Get the team together!

RICKY: Yeah. Will do, Top.

(RICKY starts to leave.)

DANNY: Hey. You still with me?

RICKY: Always. But, you gotta know something...

DANNY: What's that, man?

RICKY: *(more Operator 837 than Ricky)* You're running out of time to let go.

DANNY: What?!

(Short radio squelch then a sudden barrage of gunfire. RICKY drops dead. DANNY rushes to RICKY's side.)

DANNY: Ricky! Ricky!! Stay with me, Pal! You stay with me. Chopper's inbound, buddy. Come on, Ricky! RICKY!

RADIO (SFX): Aranya 03, this is Bushmaster 35... what is your status, over?

DANNY: *(into his radio headset)* Bushmaster 35, this is Aranya 03. Need immediate medevac support! We have one EAGLE DOWN! Roster #023! Over.

RADIO (SFX): What is status of Target, over?

DANNY: What?! Dry Hole. I repeat, Dry Hole. Wali's gone. Out.

(TIC Light off. We hear a chopper inbound. DANNY is trembling with rage and loss. He sits in his guilt. He pulls his pistol from his holster. He looks at it. He moves the muzzle underneath his chin.)

DANNY: I'm sorry.

(KENNY leaps into action and waves both arms to signal a squelch.)

Scene 15

Short radio squelch. RICKY rises and exits as lights shift to MALIK JAN who is calmly sitting cross-legged.

MALIK JAN: Salam, Mr. Danny. Sit.

(DANNY glares at MALIK JAN and remains standing.)

DANNY: Malik Jan.

MALIK JAN: I am very sorry to hear about Sergeant Ricky.

DANNY: You already know? Ah. Of course you do.

MALIK JAN: He was an honorable man. Said Wali's evil knows no end.

DANNY: As a tribal elder, you also know honor demands that I avenge my Teammate's death... wherever that leads me.

MALIK JAN: Yes, I do understand that.

DANNY: Why won't you fight alongside us to do what's right?

MALIK JAN: Mr. Danny, who determines what's right for us? Do you think Said Wali's men are the only ones who pull us out of our homes? Do you not do the same to us when you are looking for them? Do you know how many foreign armies have come through this land seeking the same thing you seek—to change it... change our people?

DANNY: More than a few.

MALIK JAN: Over the centuries... thousands! Our poems and our tribal songs are all about you and those who came before you. You shatter upon our rocks like a small ship being tossed about on the waves. Those rocks are faith, honor, and family. Immovable. But do you really want to know why you will never succeed here, Mr. Danny?

DANNY: Why is that?

MALIK JAN: You Americans have all the technology, all the wealth, and all the weaponry at your disposal. But Afghans have the greatest advantage of all... Misery. Our people endure suffering for years that no one else in the world can stomach for even a few minutes. There. You see? You just looked at your watch.

DANNY: What does that have to do with anything?

MALIK JAN: Everything. You have other pressing engagements. You have no time for this conversation. You must be behind your large walls before the sun sets and Said Wali is about. Yes, you Americans may have many watches Mr. Danny, but we have all the time...

(SFX: short squelch. We hear Christmas music, and then a Skype call ringing.)

DANNY: Well, this was fun. But I have to go.

MALIK JAN: Of course, you do.

Scene 16

DANNY rushes to a laptop. LYNN appears and also faces a laptop.

DANNY: Hey, baby! Damn. You are a sight for sore eyes.

LYNN: Hey there, Green Beret. You being safe over there?

DANNY: Of course. *(hides weapon from view)* You know, paperwork stuff.

LYNN: Yeah, right.

DANNY: I'm serious. Besides, you got your lucky rabbit's foot and all your healing oils to keep me safe, right?

LYNN: Watch your mouth, Patton. It's worked for 15 years, hasn't it?

DANNY: Yeah... I suppose. Wow. Look at all the Christmas decorations.

LYNN: You like them?

DANNY: Hell yeah. Looks great.

LYNN: Did you get your Christmas packages?

DANNY: Not yet.

LYNN: God bless the military mail! Well let me know when you get them. Your Mom mailed some stuff for you too.

DANNY: Sounds good. You okay?

LYNN: Sure. Nothing quite like a Skype Christmas... again.

DANNY: Yeah.

LYNN: We just miss you, baby. You doing okay?

DANNY: I'm good. Same shit, different day. Seeing you makes it all good though. I'd kinda like to see more!

LYNN: Shhh... Caiden's right over there, you perv. Later.

DANNY: Deal! Hey, how was Caiden's Christmas play?

LYNN: Here, ask him. He wants to sing you his part from the program, OK? So be prepared. Caiden, come here and sing for Dad.

DANNY: That's a great idea! Hey Possum, come over here and sing me that beautiful song... *(warbling)* mi-mi-mi-mi!

(OPERATOR 837, frustrated with DANNY'S lack of progress, usurps KENNY'S authority and signals

Valhalla. Suddenly the TIC light flashes and alarms start wailing. Power shuts off and their call is dropped. DANNY and LYNN scream together.)

DANNY: Lynn?! No!!!

LYNN: Danny?! No!!!

DANNY: What the fuck?! Wali, you motherfucker! Okay, asshole. Let's PLAY!

(DANNY grabs his rifle and moves across stage in an assault. He is suddenly gripped in an agonizing headache and nightmare.)

KENNY: Mach 3...

(KENNY, furious at 837, musters a contingency action and signals the OPERATORS to conjure the assault into a night terror.)

KENNY: Swarm!

(The OPERATORS move menacingly around DANNY, just out of his reach.)

KENNY: Halt!

They are demon enemy combatants, tormenting him. DANNY is amped up in combat mode, but he can't track them, can't get a shot off, can't fight back.)

KENNY: Get him.

OPERATOR 599: You said you wouldn't leave me! You said you'd come home!

OPERATOR 726: You have to miss my birthday again?!

OPERATOR 837: You rat me out, take my team away from me?! I'm gonna get you, Patton!

KENNY: Let go!

OPERATOR 595: How many of us die in the process, Top?!

DANNY: Ricky! Ricky!!

(TIC light off and alarms out. OPERATORS exit. LYNN shakes DANNY's shoulder.)

LYNN: Danny!

(DANNY screams, lunges away, and points his weapon at LYNN. She recoils to the other side of the stage, terrified.)

LYNN: Danny please!

(The sound of a vortex sucking DANNY into darkness as everything and everyone but LYNN disappears. DANNY comes to. He is horrified at himself.)

DANNY: Oh god. Oh my god.

(Sudden blackout.)

END OF ACT ONE

LAST OUT

ACT TWO

Scene 1

Music takes us into darkness. Then, in the darkness, wailing alarms overtake the music. The TIC light spins as actors get into place: DANNY in his nightmare with the OPERATORS circled around him. A radio squelch and immediate lights up.

KENNY: Let go!

OPERATOR 595: How many of us die in the process, Top?!

DANNY: Ricky! Ricky!

(TIC light off and alarms out. OPERATORS exit. LYNN shakes DANNY'S shoulder.)

LYNN: Danny!

(DANNY screams, lunges away, and points his weapon at LYNN. She recoils to the other side of the stage, terrified.)

LYNN: Danny, please!

(DANNY comes to. He is horrified at himself.)

DANNY: Oh god. Oh my god.

(DANNY rushes to LYNN and picks her up off the floor.)

DANNY: I'm sorry, I'm sorry, I'm sorry.

LYNN: Baby it's OK, you're home. You're *home*.

DANNY: I'm sorry I'm sorry I'm sorry.

LYNN: Shhhhhh. Oh, sweetie. Come here. Sit down. You're sweating so bad. Nightmares again? The sheets are soaking wet.

DANNY: I know. That's why I came out here on the couch. I'm sorry, baby.

LYNN: I know. I know.

(She holds him. Takes a breath.)

LYNN: Before you redeployed, I went to a Steel Magnolias retreat. There was this group there, The Heroes Journey. They said they can help with... with what you're going through. *(holds up a brochure)* Whaddya think?

DANNY: *(moving away)* I think that will get me pulled off the deployment. Maybe when I get back.

LYNN: Promise me.

DANNY: We'll see.

LYNN: Danny. We gotta do something. You're grinding your molars into dust.

DANNY: I'm good. Really.

LYNN: Prove it.

(LYNN pushes the brochure into his chest and leaves.)

Scene 2

Radio squelch. Slam to Afghanistan.

MALIK JAN: Mr. Danny!

DANNY: *(sneering)* Malik Jan. What do you want?

MALIK JAN: He took them. He took our children!

DANNY: Who?

MALIK JAN: Said Wali. At dusk. Just outside the mosque. He took them. Wali left a letter at the mosque steps for me to read to the village.

(MALIK JAN pulls out the letter and gives it to DANNY.)

DANNY: *(reading the letter out loud)* "Malik Jan. You have betrayed your God and your people. You collaborate with the American Dogs at every turn. The children of Kara Khel die at sunrise."

(handing the letter back to Malik Jan) Have you gone to the District Governor?

MALIK JAN: Yes. He laughed at us. He says, "This is what we get for sleeping with devil insurgents at night and helping greedy Americans during the day." Mr. Danny. You are our only hope!

DANNY: Malik Jan. I am truly sorry. But I don't know where they are. And if we start searching buildings, Wali will kill those kids. *(turns to leave)* I'm sorry.

MALIK JAN: He'll also kill them if we do nothing! *(takes a breath)* Mr. Danny. We have not been good hosts to you. I have dishonored you. For that, I am truly sorry. But these are our children. Without them, our village has nothing. No future. Help us, and I will stand with you for eternity.

(A beat as they look at each other.)

DANNY: Hey Freddy!

(A beat. OPERATOR 595 nudges OPERATOR 726.)

OPERATOR 595: Get in there, kid. Keep it tight.

OPERATOR 726: Yes sir!

(OPERATOR 726 steps into the scene as FREDDY.)

FREDDY: Yeah, Top?

DANNY: Tell the interpreter to crank the generator and get the broadcast station ready.

FREDDY: Roger that.

MALIK JAN: Broadcast? Is this... is this a joke?

DANNY: No sir. It's a RIAB, a Radio in a Box. Wali expects us to come blindly searching for your kids, right?

MALIK JAN: I suppose... yes.

DANNY: Well, he doesn't expect us to show up in his rat's nest... from the inside out. Remember those hand-crank radios we handed out on patrol to your people?

MALIK JAN: Of course I do, yes.

DANNY: My guess is shitheads listen to the radio too! Let's go, Malik Jan. It's showtime. *(Danny takes the mic)* This message is for the men who serve Said Wali. This is Danny Patton. I am here with the people of Kara Khel Village. They are here crying for their children. And where are you? Hiding in some hole after taking these innocent boys whose only crime was to pray to the same God you worship. You are not Wali. You can save your own life while he isn't watching. Let those children go. Or your own fathers will kill you slowly, while we watch.

(DANNY hands the mic back to FREDDY, who exits looking impressed.)

MALIK JAN: Manana. Thank you. I am in your service.

DANNY: Staray Mashay. I am in yours. But Said Wali is not going to be happy about this. He will return with a vengeance... for your people, and mine.

(MALIK JAN'S gratitude gives way to a cold, calm, righteous fury.)

MALIK JAN: You are right, Mr. Danny. But honor is a curious thing. Young Said Wali is not the only one capable of anger. I believe the wind has changed direction. Do you feel it?

DANNY: Yes. Yes. I think I do.

(MALIK JAN exits as KENNY enters.)

KENNY: Mission accomplished. You got into the village. *(moves to the Wall of Honor)*

DANNY: Yeah. I guess. *(follows to the Wall)*

KENNY: All those kids. Must've been intense.

DANNY: Not really. We knew the risk and we took it.

KENNY: Exactly the same with these guys. *(pointing to the many memory bracelets on the Wall)* These guys came to the team after I left?

DANNY: After the war started.

KENNY: That's a lot of guys to lose off one team.

DANNY: Careful. Those memory bracelets are all I have left of them.

(Note: the names and causes of death of the following fallen warriors can be adjusted to honor brothers and sisters from your cast's life experiences.)

KENNY: Victor?

DANNY: Ambush.

KENNY: EZ?

DANNY: Shot at close range.

KENNY: Brad?

DANNY: Found him dead in a hotel room. Guess he couldn't take any more.

KENNY: How many more of these bracelets is Wali worth?

DANNY: He's not worth ANY of them!

KENNY: So let go of this shit and come with me, man. Please!

DANNY: All these guys lost and you want me to just let it go?

KENNY: Yes. I do, Danny. This isn't your fault, man.

DANNY: Doesn't matter. I gotta set this shit right. There is no other way out.

Scene 3

Radio squelch. Slam home. DANNY is momentarily disoriented.

LYNN: When did you get home?

DANNY: Home? Where's Caiden?

LYNN: He's out riding his bike that I got him for his birthday... last week.

DANNY: Great! Didn't you tell him our plane broke down in Germany?!

LYNN: He doesn't understand that and he doesn't understand why you keep leaving us.

(DANNY turns and stares at the Wall of Honor.)

LYNN: All you do is come down here and stare at this wall. I'm not used to seeing you like this.

DANNY: Well, I'm not used to feeling like this.

LYNN: And I bet you didn't call The Heroes Journey like you promised you would, did you?

DANNY: I didn't promise that. If I get diagnosed with that PTS shit, they're going to throw me off the team. I have a team and an entire village to worry about.

LYNN: Oh, I know.

DANNY: You know what? I don't think you have a fucking clue what it's like fighting this war day after day!

LYNN: I don't have a fucking clue, huh?!? I think that *you* don't have a fucking clue. Who do you think picks up the pieces of what you boys break over there?! Who sprinted to Ricky's house and gathered up his kids as Linda fell to her knees when the Chaplain told her the news? Huh?! Same for EZ. Same for Victor. Every time you have an Eagle Down... I'm back here trying to put the nest back in the tree.

(A stunned beat.)

DANNY: I'm... I'm sorry, baby.

LYNN: Are you?

Scene 4

Radio squelch and slam to Afghanistan. MALIK JAN approaches with his right hand over his heart in greeting.

MALIK JAN: Salem Alekum, Mr. Danny!

DANNY: *(Pashtun Greeting)* Salem, Malik Jan!

MALIK JAN: Thank you for walking home with me, my son. But it really isn't necessary.

DANNY: These are dangerous times, old friend. How was your visit to Sarwah Village today? You went to see your cousin, right?

MALIK JAN: Yes, I did. Everyone there is talking about the rescue of our children. They want to make a stand against Said Wali there also.

DANNY: I know. They are asking for Green Berets to live in their village.

MALIK JAN: Congratulations, Mr. Danny.

DANNY: You and your people did it... not us.

MALIK JAN: I suppose.

DANNY: Something's bothering you. What is it?

MALIK JAN: I fear it is too late, my son. It's too late to make a stand.

DANNY: Why would you say that?

MALIK JAN: I heard your politicians say that you will all leave Afghanistan very soon. When you do, the Taliban will take a heavy price from my people for working with you.

DANNY: I heard that as well. We've got to move fast. I wish we had more time.

MALIK JAN: A wise elder once said, "Ours is not to question why, but to make the most of the time we have left."

DANNY: That's heavy. *(beat)* Wait... Gandalf said that to Frodo in *Lord of the Rings*!

MALIK JAN: *(smiling wryly)* Was he not an elder, Mr. Danny? *(more somber)* Your friendship has meant many things, my son. Including the chance to see some very amazing films with my grandson. Thank you.

DANNY: It is my honor, sir. *(takes a green chemlite out of his pocket)* Here. It's not much. Give this to your grandson, Fahim.

MALIK JAN: I've seen these. These are light sticks?

DANNY: Yep.

MALIK JAN: How do you use them?

DANNY: You just break it, shake it... and throw it.

(DANNY hands the unwrapped chemlite to MALIK JAN.)

MALIK JAN: And what happens when it burns out?

DANNY: Easy. Just leave it where it is. They're cheap. Easy to replace. Get another one. We've got thousands of them.

MALIK JAN: *(stares reflectively at the chemlite for a beat)* Sometimes, I feel like this light stick.

(They stare at the chemlite together.)

DANNY: That makes two of us, old friend.

(They glance at each other briefly, as if knowing it's their last time together.)

MALIK JAN: You are returning home soon?

DANNY: In a few days. But I'll make sure that the new team doubles down on protection of your family.

MALIK JAN: I've lived a long life. But I worry about my wife and my grandson.

DANNY: You have my cell number. I'll be back as soon as I can, so that we can finish this thing. I promise.

(Radio squelch. Lights shift. MALIK JAN disappears as KENNY approaches DANNY.)

KENNY: It's time for you to go home, man. Think about what this is doing to your family.

DANNY: I need to be over here, Kenny. My family is fine!

KENNY: You think so?! *(disengages from Danny and speaks to OPERATOR 599)* Show him the war he didn't see. *(to OPERATOR 726)* Work the edges, kid. Just the edges.

OPERATOR 726: *(frustrated)* Yes sir.

(Radio squelch.)

Scene 5

OPERATOR 599 becomes LYNN nervously watching a news bulletin. DANNY watches, unseen by her. We hear a dog barking offstage, and a phone ringing. The news reports a chopper crash, and LYNN watches in growing horror. Note: pre-recorded news bulletins are available as part of licensing.

CAIDEN: *(from off-stage)* Mom, have you seen my basketball shoes?

LYNN: They're in the closet, baby.

(The news broadcast continues, growing in volume. The dog barks and barks. The phone rings and rings. The noises and the news are overwhelming. Lights tighten on LYNN. She starts to come undone.)

CAIDEN: *(from off-stage)* Mom we gotta go! The game is in five minutes. You taking me or not?

(LYNN desperately tries to hold it together. Lights tighten even further. There's a frantic knocking at the door.)

CAIDEN: *(from off-stage)* Mom, Ricky's wife Linda is here. She looks upset. Does this mean we're gonna be late again?

(LYNN breaks. She screams. She is ripped apart by her scream. Her scream smashes all the noise until all we hear is her scream. When at last her scream has finished, lights and sound restore to normal levels. She is empty. She takes an unsettling moment to restore herself to a carefully constructed calm.)

CAIDEN: *(from off-stage)* Mom??

LYNN: *(plastering on a fake smile)* I'm on my way, baby.

(DANNY is shell-shocked by what he has just witnessed.)

DANNY: I... I didn't know...

KENNY: *(unyielding)* Yeah. You didn't.

(The erratic heartbeat fades in. DANNY staggers a few steps away. The OPERATORS drop character and speak to KENNY.)

OPERATOR 599: Sir, his heartbeat. It's fading fast.

OPERATOR 837: Which means it's time to abort this mission.

KENNY: No!

OPERATOR 595: Sir, if we stay any longer, we will descend to Hell with him.

KENNY: He's on the edge. We still have time. One more push. *(beat)* Take it to Protocol Level Charlie and introduce him to his new boss!

(OPERATORS nod and move into action.)

Scene 6

Radio squelch and heartbeat out. Slam to SMITH'S office in Fort Bragg. SMITH sits behind his field desk.

SMITH: Report, Patton!

DANNY: Lieutenant Colonel Smith?? Ain't karma a bitch!

SMITH: No salute for your new battalion commander?

DANNY: *(casually salutes)* Congratulations on your promotion and command, Sir. It's good to see the officer management system operating without undue external influence.

SMITH: Tread lightly, Patton. As of now, your mission is changing. The President is going to pull us out soon. We're shutting your village bullshit down.

DANNY: Shutting it down? Sir, our strategy is working. I promised those people in Kara Khel if they stood up against Said Wali, we'd stay with them for the long haul.

SMITH: Oh, I didn't realize that you had promised. Tell you what, if I get the chance to brief the President tomorrow I'll make sure to tell him you disapprove! Meanwhile, here in the real world, there will be no, I repeat, NO living in villages. You'll live on the firebase, train the Afghans behind the wall, and send them out to do their own fucking dying.

DANNY: Sir, we've bled for those villages.

SMITH: Yes, we have.

DANNY: Those villagers are the only chance we've got. Do you have any idea how many men I've lost doing this mission?

SMITH: We've all lost, Patton.

DANNY: This is bullshit.

SMITH: Better watch yourself, Patton. I don't buy your Lawrence of Arabia bullshit. Never have. You do know that you are a freak-show around here, right?

DANNY: They're going to be murdered, Sir!

SMITH: Sergeant. You are delusional. See, people die in war. Right? Happens every day. This is no different.

DANNY: Sir, just let me warn them. They don't know this is coming. They don't know that we're doing this.

SMITH: This conversation—

DANNY: Just let me warn them, okay? I just need to let them—

SMITH: Is over! Sergeant, you will have no contact with anyone from that village. You and your team will stay out of that village unless you obtain permission explicitly from me. Otherwise, I'll have you court-martialed for disobeying a direct order. We clear?

DANNY: We've got blood on our hands, Colonel. You know that, right?

SMITH: Not me, Patton. *(shows hands to DANNY)* See? Dismissed.

Scene 7

Radio squelch. LYNN rockets toward DANNY.

LYNN: Danny Thomas Patton! What the hell did you do?

DANNY: What are you talking about?

LYNN: Caiden is upstairs crying into his pillow. What the hell did you say to him this time?

DANNY: He told me he was gonna skip college and join the Army. He's not old enough to decide that.

LYNN: What are you talking about? You were younger than he is now when you joined up!

DANNY: I don't care. He's not old enough to make that kind of decision. He has no business being in the army. And besides, you and I agreed he's going to play basketball and he's going to be an engineer.

LYNN: We agreed, huh?

DANNY: That's right.

LYNN: Since when do we get to decide what he wants to do with his life? Huh? What is this really about?

DANNY: *(holds up a memory bracelet)* I have 23 of these bracelets. I don't want Caiden to be 24. Besides, these assholes in charge just want to shut it down in Afghanistan, do it again somewhere else. All those guys that fell? For what?

LYNN: Your friends died doing exactly what they loved, because they're Green Berets.

DANNY: Oh come on! Look around, Lynn. People back home don't give a shit. They don't know what we gave up.

LYNN: I know.

DANNY: When your dad and my dad came home from Vietnam, they threw dog shit at them at the airport.

LYNN: I know.

DANNY: Now they thank me for my service. Then walk straight out the front door and beat their neighbor over the head with a fucking ax handle. What's the point, huh? Dying for your country is one thing, and I am fully prepared to do that, but I have no idea how I'm supposed to live for it. Everything we bled for over there. So they can just tear each other apart right here at home. How long do you think it's going to be before these dingy yellow ribbons turn back into dog shit? Except this time they're going to be throwing it at Caiden. And I'm not putting my son through that. He wants to do this because of me. I'm going to fix that. When I get back from this deployment, I'm done.

LYNN: You're done?

DANNY: That's right.

LYNN: Do you remember the night you proposed to me?

DANNY: Of course.

LYNN: You said if you stopped loving it, we could get off the ride... together. So when exactly did you plan to include me in this epiphany of yours? Huh?! Every time you deploy, I walk around with my gut tied in a fucking knot, petrified whenever there's a knock at the door. Nineteen years, being Mom AND Dad, an empty seat next to me at Caiden's games and awards ceremonies.

DANNY: You think I don't know that?!

LYNN: You can't get those moments back. Sometimes when I have to sing him to sleep because he's crying for you so loud, I think my heart might actually break in two and fall on the floor. I've had thousands, *thousands* of lonely nights, just daydreaming of you and me slow dancing like we used to. Back when you smiled a lot.

DANNY: I still smile.

LYNN: No you don't! I did all of that because you told me it was worth it! And now you're telling me it's not?

DANNY: It's worth it. It *is* worth it. If we can keep Caiden away from the military. If we can keep him away from this fucking war. If he could have a real career, get married, make you a grandmother.

LYNN: How dare you.

DANNY: I'm too far down this road, baby. But not Caiden. He still has a dream.

LYNN: His dream is to be like his daddy! To make you proud.

DANNY: Proud?!

LYNN: Yes.

DANNY: What, by going off and dying for a bunch of fucking people at home who hate each other? It doesn't make me proud. That makes me sick. And when I get home, I'm done.

LYNN: You're done. Well, if you are done, I couldn't be happier. But after everything that we have been through, I don't want you back here if you're just giving up.

DANNY: Tell you what, I'm going back to Afghanistan tomorrow, where shit makes a little bit more sense. But before I do, I'm going to tell you, like I told that boy, he's not joining up because I won't allow it.

LYNN: He is going to do exactly what he wants, because that is exactly what *you* do. So you better get your head out of your ass, Danny Patton!

(LYNN exits.)

DANNY: I'll see ya when I see ya... *(softly, to himself)* but if I don't...

Scene 8

Radio squelch. Slam back to SMITH's office in Afghanistan.

SMITH: You're done, Patton.

DANNY: Excuse me, Sir?

SMITH: Cut the shit, Sergeant! I know that you left your firebase last night. We've got the whole thing recorded on the cameras outside the gate.

DANNY: Colonel Smith, it was just seven kilometers to Kara Khel Village, and right back.

SMITH: I don't give a fuck if you drove seven feet! I told you that you were to gain permission from this Command first.

DANNY: Sir. Yes, sir. I understand. But Malik Jan called me on my cell phone, Sir. He was in trouble. He was begging me to help him. I—

SMITH: Oh, I didn't know that he called you. I don't give a fuck.

DANNY: He was murdered because I didn't help him.

SMITH: He got what he deserved. And I said, I don't give a fuck.

DANNY: Well you should give a fuck. You can't make a promise that gives people hope and then abandon them when it becomes politically inconvenient.

SMITH: No, no. Don't lecture me on how to work with the savages, Patton. It's clear that you're starting to forget what side you're supposed to be on.

DANNY: Sir. I'm clear what side I'm on. But those so-called savages are the only shot we've got left.

SMITH: The only shot we've ever really had is if we march every one of these Taliban fucks down the hill and put a bullet in their fucking skull cap.

DANNY: Sir, Malik Jan wasn't a Taliban.

SMITH: Bullshit. He was just a different shade of bad! Do you understand me?

DANNY: I understand perfectly.

SMITH: Special forces is changing and we do not have any room on our teams for undisciplined renegades like you. You and your guys look like shit. Beards out of regs, man-jammies on, ball caps on patrols, and you're wearing bullshit kid's jewelry like that fucking green bracelet that I've told you to take off how many times?

DANNY: Sir, Caiden gave me this—

SMITH: Who the fuck is Caiden?

DANNY: *(gritting his teeth)* He's my son.

SMITH: I told you that I was going to get you. And now, now I have you. I'm issuing you a letter of reprimand and I'm pulling you and your team out of the field effective now.

DANNY: You can't do that!

SMITH: You have no idea what I can do to you—*and your team*—right now.

DANNY: Sir, it was just me. All right? My team had nothing to do with it. They didn't know I left the wire. I went out by myself.

SMITH: Bullshit! Don't fucking lie to me. You expect me to believe that you left the firebase at night on your own in hostile territory?

DANNY: I've been running around this hostile territory on my own for ten years, at night. So yes, I do expect you to believe that because it's the truth. Check your video camera, Sir, I'm not lying to you. It was just me.

SMITH: Fine. But I gotta show the chain of command this is being dealt with.

DANNY: Come on, Sir. My team doesn't deserve this.

SMITH: Well, there is a way that you can save your precious team and keep all this quiet. You are going to come out of the field and work in this headquarters, directly for me. You will accept the letter of reprimand and a substandard evaluation report. So I don't catch any shit from your little senior officer fan club at headquarters.

DANNY: That'll ruin my career, Sir.

SMITH: I thought you didn't sign up for a career, Patton. I thought you signed up to be a Green Beret. Here is your one chance to keep it.

(DANNY aches but signs the papers SMITH is holding out.)

SMITH: Right there, too.

(DANNY signs.)

SMITH: Done. Get back to your firebase and give your team their goodbye kisses. I want you on the resupply chopper in three days and you're reporting into the First Sergeant. Nice and easy.

DANNY: Yes, Sir.

(DANNY starts to walk away.)

SMITH: Hey, one more thing. You're going to be working for me now. And in this headquarters, we have fucking standards. Why don't you start by acting like a grown-ass man for a change?

DANNY: Yes, Sir.

(DANNY again starts to walk away.)

SMITH: Let's start right now. Take the bracelet off and put it in my trash can.

DANNY: Sir, please listen, my son Caiden gave me this—

SMITH: I am tired of fucking around. This is about respect. Throw it in the fucking trash or get your team in here and I'll destroy every one of your careers!

DANNY: Shut the fuck up! Now you listen to me, junior, and you listen real Goddamn good. This here is a magic silly band and it's expensive as fuck to take it off this wrist. A lot of men have tried, and something tells me you're not quite up to the task. So if it's all the same to you, I'm going to get back to my team now.

(They glare at each other. SMITH blinks first.)

SMITH: Salute me first, then.

DANNY: *(with a perfect salute)* You have a nice day, sir.

Scene 9

We hear the same explosion we heard at top of show. The TIC light spins and splashes its red light across the stage throughout the scene. A very slow and ominous heartbeat thuds underneath. SMITH disappears and KENNY enters. DANNY is shocked.

DANNY: *That's* when I bit it? On the drive back to the firebase after meeting with Smith? That's why I wasn't paying attention to the road?!

KENNY: Yeah, man.

DANNY: Are you fucking kidding me?

KENNY: Hey. Saved you from working with that asshole.

DANNY: Fuck you. The last time I talked to my family, I was so angry. I was so pissed. My letter, number 710, my letter to Lynn and Caiden...

KENNY: What about it?

DANNY: *(heartbroken and ashamed)* I didn't write it before I left, man. I was so fucking pissed. How could I do that?

KENNY: That's right, Danny. There's no letter. There's no letter to Lynn. There's no letter to your son.

DANNY: *(starting to unhinge)* Fuck man, they're going to hate me.

KENNY: It doesn't have to be like that.

DANNY: How can I let them go? When I don't know they're going to be okay?

KENNY: Look, Danny, none of us really know that. You just got to let them go.

DANNY: I can't, I can't, I can't.

KENNY: Yes, you can. Look, Danny, look!

(KENNY points across the stage. Lights shifts. LYNN and CAIDEN appear. CAIDEN is headed to basic training. They silently say goodbye, embrace, etc., underneath the following.)

KENNY: They're going to be okay.

LYNN: *(pulls out a letter and reads as though composing it)* My dearest Danny boy, today is the day that I have feared since you hawked that microwave and got Caiden his first uniform and that wooden rifle. He joined up, just like he said he would. I just dropped him off at the airfield. So proud. So strong. So stubborn. Like father, like son. People keep telling me how lucky I am to have something that makes saying goodbye so hard. What a crock of shit that is, but I don't want you to worry about us. Okay? We'll find a way through this. And you will always have my heart.

(A cell phone rings. Lights shift. MALIK JAN is crouched in his bedroom, whispering into his phone.)

MALIK JAN: Danny?

DANNY: Malik Jan?!

MALIK JAN: Said Wali is here. He's coming to the bedroom now. I need you to come for my grandson. He will be in the special hiding place, you and I discussed. You remember?

DANNY: I remember...

MALIK JAN: We knew this was my fate. Thank you for reminding me of who I truly am. Always remember—

(MALIK JAN disappears as we hear a spray of bullets and men yelling.)

DANNY: Malik Jan! NO! *(beat)* I'm sorry...

(Lights shift. CAIDEN sits beside LYNN, who is presented a folded flag by the FUNERAL PRESIDING OFFICER.)

FUNERAL PRESIDING OFFICER: On behalf of a grateful nation. *(salutes)*

("Amazing Grace" plays on bagpipes; either a live piper or SFX. LYNN moves to the Wall of Honor and places the folded flag on top of the Wall.)

LYNN: And even though you are gone, I will wait as long as it takes, Master Sergeant Patton.

DANNY: I can't do this without you. I can't even breathe without you.

Scene 10

The OPERATORS turn off the TIC light. The slow, ominous heartbeat continues underneath.

OPERATOR 599: Sir, we are at the point of no return.

OPERATOR 837: We are initiating the Valhalla Abort Protocol now, Sir.

KENNY: We're so close!

ODIN (SFX): (*garbled radio transmission*) $%&*! – Valhalla! $%&*! – Valhalla!

(KENNY pulls out his walkie-talkie.)

KENNY: This is Suggins.

ODIN (SFX): *Garbled radio transmission*

KENNY: Sir, I just need a little more time! We've almost got him!

ODIN (SFX): *Garbled radio transmission*

(OPERATORS 726 and 595 approach KENNY.)

OPERATOR 726: Sorry, Sir. He was a good man.

KENNY: Is. He IS a good man.

OPERATOR 595: We'll see you back up topside, sir. Let's go, kid.

OPERATOR 726: *(reluctant to leave)* But...

OPERATOR 595: Let's go!

(OPERATORS exit.)

ODIN (SFX): *Garbled radio transmission*

KENNY: *(back into walkie-talkie)* With all due respect, Sir, fuck you. I will not leave him behind, even if it means I go down with him. Suggins out. Master Sergeant Patton, report. Front and center. Right now!

DANNY: Why are you doing this to me, Kenny?

KENNY: I'm not doing a damn thing. Look, you have to let go. You have to do it now. If not, you're done.

DANNY: I'M NOT DONE! Wali is still out there, Malik Jan is gunned down, Lynn is alone, and Caiden is going down the same bullshit path I did. How can I be done?!

KENNY: You hold on there. You think you're the only one? The lone warrior who ain't done yet? There's a whole bunch of combat boys and girls, including me, felt the same shit you're feeling right now. Wondering what it was all for. But it's harder than any life that you'll have to take or any mission that you'll have to leave. *Letting go.* It is the only way to know peace.

DANNY: Peace?! You don't get it, do you, Kenny?! How am I supposed to let my son fight a war... *(he slams his chest with his fist)* ...that I COULDN'T FUCKING FINISH?! How can you expect me to die like this?! There's no peace in that. Knowing that everything I did was worthless!!!

(TIC light starts to flash. The heartbeat gets louder, uneven. Hell's alarm wails. Banyons—enforcers released from Hell to reclaim renegade souls who don't let go—are coming. KENNY knows time is short. KENNY yells at ODIN.)

KENNY: Wait!!!

(KENNY grabs DANNY.)

KENNY: Danny, you got to hear me now, man. Your heart is about to stop beating. Look, that alarm, that's for you. They're not going to let me stay, man. Darkness is coming. There is never, ever enough time.

(DANNY prepares for his final stand-alone.)

DANNY: Go to hell, Kenny.

KENNY: Going there is a choice too, Ranger Buddy. And everyone makes it on their own.

(We hear more garbled radio transmissions. The heartbeat skips.)

KENNY: *(into his walkie-talkie)* Goddammit, sir, I'm not leaving him!

(But the heartbeat flutters, and darkness swallows KENNY, leaving DANNY alone.)

Scene 11

The Wall of Honor disappears. All that is left is the spinning TIC light and DANNY, clutching his rifle, assaulted by sound: Hell's alarm, distant gunfire, garbled bar music, and his own fluttering heartbeat. DANNY points his rifle in every direction, but there's nothing and no one to target. All sounds fade. We have descended into Hell. The TIC light flashes in silence.

DANNY: Lynn?

(Banyons shriek.)

DANNY: Caiden?

(Banyons shriek.)

DANNY: Anybody?!

(The following voices all come from off-stage, and from various directions. Note: these are names of fallen warriors and all except Patton's can be substituted with names of fallen warriors from your own community.)

ROLL CALL PRESIDING OFFICER: *(from off-stage)* Call the roll! Captain Robinson.

CAPTAIN ROBINSON'S VOICE: Here!

ROLL CALL PRESIDING OFFICER: Staff Sergeant Johnson.

STAFF SERGEANT JOHNSON'S VOICE: Here!

ROLL CALL PRESIDING OFFICER: Sergeant First Class Cervantes.

SERGEANT FIRST CLASS CERVANTES'S VOICE: Here!

ROLL CALL PRESIDING OFFICER: Patton.

DANNY: Here!

ROLL CALL RESIDING OFFICER: *(from off-stage)* Master Sergeant Patton.

DANNY: Right here!

ROLL CALL PRESIDING OFFICER: *(from off-stage)* Master Sergeant Danny Patton.

DANNY: I said I'm right here, motherfucker!

(A 21 Gun salute echoes. The first volley hits DANNY violently.)

DANNY: Fuck! Stop!

(The second volley hits DANNY violently.)

DANNY: What are you doing?

(The third volley knocks DANNY to the ground. He climbs back to his feet, engulfed in the madness of it all.)

DANNY: Is that all you got, huh?

(DANNY holds up his weapon defiantly.)

DANNY: I am Danny Patton, and this will keep me safe!!

(DANNY suddenly loses confidence in his weapon. He tugs at his body armor.)

DANNY: I am Master Sergeant Danny Patton and THIS will keep me safe.

(But he suddenly loses confidence in the body armor as well. Lights shift as KENNY desperately crawls his way out of the darkness.)

KENNY: There's only one thing that ever kept you safe!

(DANNY regards KENNY and then notices the silly band on his wrist.)

DANNY: I am Master Sergeant Danny Patton. U.S. Army, Special Forces.

(He takes the silly band off his wrist and thrusts it into the air.)

DANNY: This will keep me safe! Do you hear me?! This will keep me safe!!

(DANNY hears a wave coming. Icy waves of Hell crash over him. Banyons are swirling on top of him. He falls to the floor, face down, unconscious. Note: as he falls, he

puts the silly band into his pocket. Hell's alarm wails. Banyons are coming. KENNY sits beside his friend.)

KENNY: I'm here, Danny. I won't leave you. No matter what comes.

(The OPERATORS re-appear, out of breath and freezing.)

KENNY: What are you doing back here?

OPERATOR 837: We don't leave anyone behind, sir.

OPERATOR 595: Besides, I hear hell is nice this time of year.

OPERATOR 599: We don't have much time. What's the call?

KENNY: He found his silly band and that bought us a little more time, but the darkness almost has him.

OPERATOR 726: I think—

OPERATOR 837: I've employed every engagement character that I could think of.

OPERATOR 726: Excuse me—

OPERATOR 599: Yeah, there isn't much left for us to do.

OPERATOR 726: Hey... Guys...

OPERATOR 595: Dude, just hold up. Okay?! Not now.

OPERATOR 726: No! You hold up. *(pause; has he overstepped?)* Sir. Those are not the things keeping him here.

OPERATOR 599: Well, then what does he want?

OPERATOR 726: What all of us wanted before we ascended.

OPERATOR 837: To say goodbye?

OPERATOR 599: Last words.

KENNY: I'll be damned. Okay, kid. You're in. This is our last shot and we're out of time. Make it count.

(OPERATORS move into position. OPERATOR 726 walks past OPERATOR 595 to lead the recovery mission.)

OPERATOR 595: Hey, kid.

(OPERATOR 726 stops and awaits the insult.)

OPERATOR 726: Yeah?

OPERATOR 595: Save him.

(OPERATOR 726 nods and moves to the Wall of Honor for one final attempt to save DANNY. Lights shift. TIC light stops.)

CAIDEN: Hey Dad. It's me.

(DANNY looks up.)

DANNY: Caiden? My soldier.

CAIDEN: I wanted to come back for one last look at our wall before I ship out for Afghanistan tomorrow.

DANNY: Afghanistan?

(CAIDEN cannot hear DANNY. He talks to the Wall of Honor.)

CAIDEN: Your funeral was amazing, Dad. I bet there were a thousand people there. Packed outside the chapel trying to just listen to the service. So many of your Brothers were there. *(beat)* Do you remember the Skype calls from Afghanistan? There weren't many, but man... I loved them. I remember one day in 7th grade, you were deployed, and you were going to call right after school so I could be on. I walked around all day so excited. I couldn't focus on anything except talking to you. I sat at the front of the bus on the way home and it took forever! When I finally got off the bus, I ran home so fast that papers were flying out of my backpack, but I didn't even care. I just wanted to hear your voice. I sprinted through the front door: "Is he on yet?!" Mom kneeled in front of me and told me you'd already called, but barely had time to say hi because you had so much stuff going on. I asked her when you'd call back. She didn't know. She just said, "Soon." *(beat)* Soon. Yeah. I'd heard that promise before. I went to my room and locked the door. I called you every curse word in the book. Then I cried into my pillow so Mom wouldn't hear me. That happened more than once, a lot actually. *(beat)* I used to blame you, Dad. But not anymore. I get it now. You hated being away from me. That's why you got so upset about me joining up. You were trying to protect me. But I couldn't protect you. I did this to you, didn't I?

DANNY: No you didn't, son. Please don't blame yourself.

CAIDEN: So, I just needed to come say... I'm sorry.

DANNY: No, I'M sorry!

CAIDEN: My insides are chewing me up. I don't really know how to do this.

DANNY: Yes, you do! You've done it every day of your life!

CAIDEN: I just wish you were here. Because I don't know how to do this without you. This is my fault. Mine. I shouldn't have put this on you before you deployed, dad. I was going to talk to you when you got back from this deployment, you were probably worried about me or distracted and you weren't paying attention to the road and that's why it got you. But now it's on me. So I promise you this, dad... I don't care what it takes or how many I have to kill. I'm going to find them and I'm going to kill them all.

DANNY: No, son! That shit will chew you up. I don't want you to end up like Smith... or like me. Kenny! Please! I made my choice, I know that now, but I'm begging you! Just once. Just once let me be there for my boy!

CAIDEN: Until Valhalla, Dad.

DANNY: Caiden!

LYNN: *(from off-stage and simultaneous with Danny above)* Caiden?

DANNY: *(spinning to look for her)* Lynn??

CAIDEN: Down here, mom.

(LYNN enters carrying a large mailing envelope.)

LYNN: You all right, baby?

DANNY: Oh my God. Lynn!

CAIDEN: *(putting on a brave face)* Yeah, I'm all right, mom.

(But CAIDEN crumbles. LYNN goes to him and wraps him up in a hug. DANNY can only watch.)

LYNN: Hey. Hey. Hey. Come here.

CAIDEN: I miss him.

LYNN: I miss him, too.

CAIDEN: *(pointing at the envelope)* What's that?

LYNN: I don't know, it just arrived from Walter Reed Hospital. It's from Steve Hill. He is... he was your daddy's medic.

(They sit together and open the envelope. They pull out a letter. CAIDEN unfolds the letter and reads.)

CAIDEN: "Lynn and Caiden. I've struggled for a couple weeks on writing this letter, trying to find the words. I'm so sorry. I lost him. I did everything I could, but it wasn't enough. I was with him when he died."

(KENNY puts his hand on DANNY'S shoulder.)

KENNY: Stevie is fighting so hard to keep you alive. Come on, Danny. One last shot.

(Lights shift. STEVIE appears exactly where we left him at the beginning of the story: working to stop the bleeding and trying to save DANNY. The erratic heartbeat underscores throughout.)

STEVIE: You guys! I've got a pulse, weak and thready! What's the ETA on the medevac?

VOICE FROM OFFSTAGE: Three minutes!

KENNY: Let go, brother.

CAIDEN: *(still reading the letter)* "He wanted to tell you something. Both of you."

DANNY:Stevie?

STEVIE: *(speaking down to where DANNY'S body would be on the ground)* Stop talking, man. It's all right, you're going to be home to your family soon.

(DANNY realizes this is his chance to write letter 710.)

DANNY: Stevie! I've seen them!

STEVIE: What, man?

DANNY: I need you to listen to me right now. I just need you to listen to me.

CAIDEN: I can't, Momma...

(LYNN gently takes the letter from CAIDEN.)

LYNN: I got it. "To Caiden..."

LYNN / DANNY: *(simultaneous)* "I'm sorry that I didn't listen to you before I left."

DANNY: I'm so damn proud of the man you've become. The truth is I was afraid. I was afraid / of what—

LYNN: —of what might happen to you. But that little silly band you gave me reminded me what really matters.

DANNY: Don't you go down the same road I did. You are not the last out. No one is ever the last out. You will have to fight. Maybe even do things that will haunt your dreams... but it's to protect the people you love. Never revenge. Don't forget that.

(LYNN hands the letter to CAIDEN. The heartbeat begins to slow.)

LYNN: Okay, Possum?

CAIDEN: I won't, Dad.

DANNY: I'm okay. I promise. Don't you be in a hurry to join me up here... by the enemy's hand, or your own. I love you.

CAIDEN: I love you. *(Caiden flips the letter over)* Mom, there's more.

DANNY: My sweet Lynn. From the first second I held you in my arms in that shitty old bar...

CAIDEN: You were in a shitty old bar, Mom?!

LYNN: *(blushing)* Yes.

DANNY: ...all I wanted was to protect you. But instead I pushed you away. Why did we stop dancing? When did I stop smiling? It wasn't supposed to end this way.

CAIDEN / DANNY: "We made a promise to get off this ride together."

DANNY: I never did, and I never will. You have loved me more fiercely than anyone ever has. I know you're not going to want to hear this...

CAIDEN: ...but I need you to move on. Find happiness again.

LYNN: I can't. I won't.

DANNY: Hey. I mean it.

CAIDEN / DANNY: "Happiness."

DANNY: And I'll see you—

LYNN / DANNY: —when I see you.

DANNY: And until then...

LYNN: Until then...

DANNY: I love you—forever, baby.

STEVIE: Anything else, man?

DANNY: Yeah. C'mere.

(Stevie leans down as though listening to the dying DANNY whispering in his ear.)

STEVIE: Yeah, man. I'll make sure he gets it. Rest easy, brother.

(Lights fade on STEVIE as he mourns. There is a moment of silent stillness. Then LYNN looks at the envelope in her lap. She realizes there is something else in there.)

LYNN: Caiden...

(She reaches into the envelope... and goes still as she feels and sees something completely unexpected. She looks up, thrilled by one more surprise from her love. She pulls out a worn, tattered silly band, and holds it out to CAIDEN. Everyone reacts to this gut punch.)

LYNN: It's your time now, baby.

CAIDEN: I love you, Mom.

(CAIDEN takes the silly band, and puts it on his wrist. As he does...)

Scene 12

Three beeps on the heart monitor... then flatline. DANNY moves toward the light as it shifts. Darkness lifts. The soft sounds of water, and breeze, and birds. DANNY looks around unsure what's happening. The OPERATORS are stunned and relieved; they have pulled DANNY from hell. OPERATOR 595 pulls OPERATOR 726 aside and slaps an operator patch on 726's shoulder.

OPERATOR 595: *(quietly)* Nice work, kid. You're an Operator now.

(The OPERATORS all acknowledge 726: nods, small fist bumps. He nods back. KENNY walks to DANNY.)

KENNY: You ready to head up now, Brother?

DANNY: I'm ready.

(The sounds grow. Chimes and bells in the distance. DANNY looks up... and is awestruck as Valhalla appears before him.)

DANNY: Valhalla!

KENNY: *(points out over the audience)* Yeah. Ricky, Billy—

DANNY: Malik Jan?!

KENNY: Yeah, they're all there waiting for you.

(DANNY takes a huge cleansing breath. Then we hear a dog barking.)

DANNY: Axel?!

KENNY: Like my Daddy always said, if there ain't dogs in heaven, I don't wanna go.

DANNY: Hey man, thanks. Thanks for sticking with me.

KENNY: That's what we do, Green Beret.

DANNY: *(including the audience)* And thank you all for letting me say goodbye to them.

KENNY: Now let's head up to that shade tree before Ricky gets my spot.

DANNY: *(smiling)* Roger that.

(DANNY steps toward Valhalla, but is stopped by the grinning OPERATORS. He senses they were—and are—his family.)

KENNY: Hey. You won't need that weapon on this mission. You won't need it ever again.

(DANNY is unsure. KENNY and the OPERATORS smile, silently encouraging him. He places his weapon upside down into the Memorial Block, hangs his dog tags on the pistol grip. He places his green beret on the buttstock of the rifle. He looks up...)

(He ascends. He is free.)

("Ballad of the Green Beret" sung by little J—included in sound package—plays us into BLACKOUT.)

END OF PLAY

Coda

This Coda can be performed during the talkback to answer the recurring question: "Was Afghanistan worth the cost we paid?"

In the dark, we hear a news broadcast about a terrorism attack and the expected US response. Explosions. Gunfire. Whistle of incoming missiles.

Lights up on CAIDEN, in war kit, quickly rolling up his parachute. He glances around at the sounds of distant combat. He is behind enemy lines and trying to stay quiet.

He doesn't notice FAHIM creeping up on him. FAHIM is an Afghan Commando in tattered fatigues and kit. He raises his rifle. He's got CAIDEN dead to rights.

CAIDEN: This is Vengeance 03. Infil complete. Will report in after link up.

FAHIM: Do not move, American!

CAIDEN: *(Mutters to himself)* Shit. *(To Fahim)* You speak English?

CAIDEN stands slowly, not lowering his weapon.

FAHIM: Of course I speak fucking English! What are you doing here? Where is your team?

CAIDEN: My name is Caiden. I'm here to—

FAHIM steps forward and presses his rifle deeply into CAIDEN's neck.

FAHIM: Where is your team and what is your target? You will tell me, one way or the other.

CAIDEN: Look, man. I'm here to help you. I—

FAHIM: Help me?! Motherfucker, I am an Afghan Commando. We trusted you. You want to help me? After what you did to us?! You left Bagram Airfield in the middle of the night with no warning. We had no air support, no MEDEVAC, we were completely exposed. We were national heroes. Now we're living like animals. *(beat, then... menacing)* Fuck you and your help. Give me your full name or I'm going to end you right now.

CAIDEN: Patton. Sergeant First Class Caiden Patton.

FAHIM lowers his rifle a few degrees.

FAHIM: Was your father named Danny?

A stunned beat as Danny turns to face his aggressor.

CAIDEN: Yeah. How the hell do you know that?!

FAHIM pulls out a weathered light stick, and tosses it in front of CAIDEN.

FAHIM: Your father knew my grandfather.

CAIDEN: What??

FAHIM: When Said Wali kidnapped me as a little boy, it was Danny Patton who freed me. When Wali killed my grandparents, Danny Patton came for me and hid me on his firebase. If not for him, I would be dead.

CAIDEN: Holy shit.

FAHIM: Your father was family. I owe him a debt.

The two men breathe together for a moment.

FAHIM: Go, Caiden, son of Danny Patton. For his honor, and the honor of my grandfather, Malik Jan, go. I pray we never meet again, for surely then one of us will be forced to kill the other.

FAHIM starts to leave.

CAIDEN: Wait. What's your name?

FAHIM: I am called Fahim.

CAIDEN: That means quick of understanding, right?

FAHIM: Yes. Though my wife thinks perhaps I am mis-named.

They both almost chuckle.

CAIDEN: Kids? You have kids?

FAHIM: Yes. Well, I did. My wife and children were executed by the Taliban when they were searching for me.

CAIDEN: I'm sorry.

FAHIM: For what? *(CAIDEN has no answer.)* We fight the wars our politicians and generals tell us to fight, do we not?

CAIDEN: I'd like to talk to you again, Fahim.

FAHIM: We shall see Caiden Patton. We shall see. Now go. The next Afghan you meet will not be so kind.

FAHIM disappears into the darkness. CAIDEN'S tension releases; he has faced death. He gasps, tears squeeze from his eyes, snot from his nose. He laughs—it's a bit hysterical. He suddenly notes the silly band on his wrist. He reaches down and picks up the light stick. He kisses it. He holds it skyward.

CAIDEN: Well holy shit, dad. Thank you.

He laughs again and exits quickly. "Blood on My Hands" by John Ondrasik plays us into blackout.

Acknowledgements

To Chris Crowe, Michael Hague, and others who helped me move through the writer's blocks to birth this story into the world.

To all the *Friends of Last Out* across the globe. You know who you are. Thanks for rooting us on and giving us the strength to continue this mission.

To Jamie Dunn for amazing work as tour manager in 2019 and getting us up and running with the Gary Sinise Foundation in 2022.

To all of our 2019 tour hosts and volunteers who brought us into your world and your community. Thank you for stepping up.

To Jason Cannon. Thank you, man. Thank for taking my crazy dream and pouring your heart and skill into me. The way you have dramatized this story is epic.

To Karl Bury. I will never forget your kindness in showing me the craft of acting while gently helping me crack open the parts of me that were needed for this scar story to work. And of course, your directing took us to a whole different place!

To Larry Moss. You are a legend. You didn't have to open your doors and heart to me, but you did. Thank you for pushing me and believing in this story, and me.

To John Ondrasik for reaching out and making the BIG connections happen. And for cheering us on from the crowd.

To Gary Sinise for inspiring me with *Tracers* and for taking this vision to the next level. Also, a big thanks to all the folks at the Gary Sinise Foundation who welcomed us into your family and put this play into the world.

To John "Wes" Wesley. Thank you for seeing the vision early on and for bringing an amazing community of industry experts together to make it happen.

To Brittnee Henry. Thanks for helping us organize and sustain the message financially.

To Kari Ellis. Thanks for your nonstop love and outreach to communities across the country, and for letting me catch you when you fell.

To Kim Seevers. You are the heartbeat of our storytelling workshops. Your love and compassion healed a lot of hears.

To our original cast: Bryan Bachman, Lenny Bruce, and Ame Livingston. You made the impossible possible and brought this timeless story to life.

Special thanks to our original Stage Manager, Bob Pritchard. Your kind heart and can-do attitude made so much possible for us. You will always be family.

To our general manager, Megan Nemanic. You stepped into the arena in the middle of the storm. You'd never done anything like this and you brought so much light to a flurry of chaos. Thank you for bringing the true heart and soul to this production and reminding us that we can be solid leaders and good humans at the same time.

To Emma Brannon and the awesome production team: Rebecca McBee, Kat West, Paul Hudson, Mark Prator, Collin Hardesty, Ozzy Giron, and Lee Anne Meeks. You all brought next level professionalism to our bottom-up project and propelled us to a whole new level.

To Fahim Fazli. Thank you for bringing your beautiful Afghan Voice to this story.

To our 2023 cast: Heather Corrigan, Bryan Bachman, Tony Nevada, Lenny Bruce, Cooper Mann, Chris Vetzel, Josh Berduo, and Gracyn Mix. Thanks for stepping into the arena and bringing the power of storytelling to educate our civilians and heal our military community. You all really brought it.

And to the casting director who brought that 2023 cast together, Michael Sanford.

Thanks to our brilliant bagpiper, John Corcoran.

Doc Diego. For holding us all in your hands.

Thanks to Jesse Torgerson, who helped all the actors find and own their breath.

To Frank Siller and Tunnel To Towers. You stepped in at just the right moment to let us honor our amazing NYC First Responders. God bless them all!

To my boys, Cody, Cooper, and Brayden. I hope I did it right by you and made you proud.

To my old buddy Dave "Doc" Phillips and Sweet Sue. There is a whole lot of Danny Patton in you. You came to the show and went for the ride despite the pain. God bless you, Pal. You will always be my mentor.

To all the amazing donors who put your hard-earned dollars into us and the telling of this story. Thanks for believing in us and giving us the airspeed to do it right!

Thanks to my Mom and Dad, Anita and Rex Mann. You kept showing up and believing each time we performed. It means the world. I'm proud to be your son.

To my brothers, Jerry Lujan, David Martin, and Greg Parsons. You were there when I first read this play. Each of you sponsored the show in your community, wrote your own checks, and picked me up when I didn't believe. You were there for the first performance and the last. Thank you, boys. I'll never forget you.

To all of my Green Beret Brothers, special operations community, and fellow veterans and your families. God bless you. This is your story. We will never stop telling it.

To Holly Higgins for all you did in the face of losing your beloved Danny in Afghanistan and to all our Families of the Fallen. We will never forget you or the sacrifice of your loved ones. They live on with us on The Wall of Honor and the stories we tell.

To my amazing wife, Monty. None of this was possible until you smiled at me through tears, and said: "Let's do this play."

To our Afghan Allies. I'm sorry we left you behind. I pray we do better next time. If you give us a chance.

To the wonderful men whom Danny Patton is based on. We will never forget you as long as stories are told.

To Brad Bynum and your beautiful family. Thank you for being my friend. Thank you for going to war with me. Thank you for trying to come home by hosting the play. I'm sorry we lost you man. We should have done better. You deserved better.

To all the veterans, first responders, families of the fallen, families of the protectors who came to this show and found a way to let go of the pain. God bless you. And if you haven't let go yet, I pray you will.

To the families of the fallen all over America, whose items are on The Wall of Honor. You are our family now. We'll never stop telling the stories of our loved ones.

To the names mentioned of our fallen mentioned in this play and those whose stories are told but their names cannot be spoken. We will always remember. We will always honor. We will share the stories until late in the night. Until Valhalla.

ABOUT THE PLAYWRIGHT

LIEUTENANT COLONEL (RETIRED) SCOTT MANN is a former U.S. Army Green Beret with multiple tours in Afghanistan. Scott is also the founder of Task Force Pineapple, a group of volunteer veterans advocating for the abandoned Afghan People. Scott is a warrior storyteller, 3 time TEDx Speaker, and the founder of Rooftop Leadership where he shares the rapport-building skills he learned in Special Forces to help today's leaders make better human connections in high-stakes, low-trust engagements. He is also the founder of the 501c3 The Heroes Journey, where he helps warriors, first responders, and their families find their voices and tell their stories in order to bridge the civil-military divide. After undergoing a difficult transition from military life and combat where he almost took his own life, Mann not only wrote *Last Out*, but also originated the role of "Danny Patton" in the world premiere production, first two tours, and film. *Last Out* sheds light on the unforeseen toll of war on our Veterans and their families, and has been seen by over 80,000 people, toured 16 cities, led to 250+ PTS interventions by trained counselors, and been performed for 75 Gold Star Families. Yet, amidst his numerous achievements, Mann holds his marriage and three grown sons as his greatest source of pride and accomplishment.

Made in United States
North Haven, CT
04 November 2026

21769178R00113